Snow and Ice Control Manual for Transportation Facilities

Construction Series
M. D. Morris, Series Editor

Godfrey
PARTNERING IN DESIGN AND CONSTRUCTION

Matyas et al.
CONSTRUCTION DISPUTE REVIEW BOARD MANUAL

Palmer et al.
CONSTRUCTION INSURANCE, BONDING, AND RISK MANAGEMENT

Peurifoy and Oberlander
FORMWORK FOR CONCRETE STRUCTURES

Rollings
GEOTECHNICAL MATERIALS IN CONSTRUCTION

Steinberg
GEOMEMBRANES AND CONTROL OF EXPANSIVE SOILS IN CONSTRUCTION

Snow and Ice Control Manual for Transportation Facilities

L. David Minsk

McGraw-Hill

New York   San Francisco   Washington, D.C.   Auckland   Bogotá
Caracas   Lisbon   London   Madrid   Mexico City   Milan
Montreal   New Delhi   San Juan   Singapore
Sydney   Tokyo   Toronto

Library of Congress Cataloging-in-Publication Data

Minsk, L. D. (L. David)
    Snow and ice control manual for transportation facilities / L. David Minsk.
        p.   cm.
    Includes bibliographical references (p.).
    1. Roads—Snow and ice control. 2. Bridges—Snow and ice control. 3. Runways (Aeronautics)—Snow and ice control. 4. Railroads—Snow protection and removal. I. Title.
TE220.5.M56   1998
625.7′63—dc21   97-48324
                                                                           CIP

McGraw-Hill

A Division of The **McGraw·Hill** Companies

Copyright © 1998 by The McGraw-Hill Companies, Inc. All rights reserved. Printed in the United States of America. Except as permitted under the United States Copyright Act of 1976, no part of this publication may be reproduced or distributed in any form or by any means, or stored in a data base or retrieval system, without the prior written permission of the publisher.

1 2 3 4 5 6 7 8 9 0   FGR/FGR   9 0 3 2 1 0 9 8

ISBN 0-07-042809-3

The sponsoring editor for this book was Larry Hager, the editing supervisor was Scott Amerman, and the production supervisor was Pamela A. Pelton. It was set in Palatino by Estelita F. Green of McGraw-Hill's Professional Book Group composition unit.

Printed and bound by Quebecor/Fairfield.

McGraw-Hill books are available at special quantity discounts to use as premiums and sales promotions, or for use in corporate training programs. For more information, please write to the Director of Special Sales, McGraw-Hill, 11 West 19th Street, New York, NY 10011. Or contact your local bookstore.

 This book is printed on recycled, acid-free paper containing a minimum of 50% recycled, de-inked fiber.

Information contained in this work has been obtained by The McGraw-Hill Companies, Inc. ("McGraw-Hill") from sources believed to be reliable. However, neither McGraw-Hill nor its authors guarantee the accuracy or completeness of any information published herein, and neither McGraw-Hill nor its authors shall be responsible for any errors, omissions, or damages arising out of use of this information. This work is published with the understanding that McGraw-Hill and its authors are supplying information but are not attempting to render engineering or other professional services. If such services are required, the assistance of an appropriate professional should be sought.

Contents

Introduction xi
Acknowledgments xvii

Chapter 1. The Basics of Snow and Ice Control   **1**

   1.1   The 10 Principles of Snow Removal: Be a Bully—Hit It while It's Down. Match Technique to the Condition   1
   1.2   The Six Principles of Ice Control: When and How to Use Chemicals; How Much to Use   7

Chapter 2. The Nature of Winter Precipitation: A Matter of Materials Handling   **11**

   2.1   Characteristics of Snow and Ice   11
   2.2   Scope of the Task of Snow and Ice Removal   11
   2.3   Snow and Ice Removal as a Materials Handling Problem   12
   2.4   The Material Snow   17
          2.4.1   Birth of a Snow Crystal   18
   2.5   The Important Properties of Snow   21
          2.5.1   Density   21
          2.5.2   Hardness (Strength)   22
          2.5.3   Compressibility   23
          2.5.4   Slush   25
          2.5.5   Cohesiveness   26
          2.5.6   Adhesiveness   26
          2.5.7   Temperature Instability   27
          2.5.8   Age Hardening   27
          2.5.9   Effects of Mechanical Agitation   28

2.5.10 Thermal Properties of Snow   28
2.5.11 Snow Is a Unique Material   30
2.6 Water and Its Forms   30
    2.6.1 Water   30
    2.6.2 Ice and Its Properties   32

Chapter 3. Snow and Ice Control Methods   41

3.1 Introduction   41
3.2 Chemical Methods   41
    3.2.1 Use of Chemicals for Snow and Ice Control   42
    3.2.2 How Chemicals Melt Ice   44
    3.2.3 Chemical Selection   48
    3.2.4 Application Methods   51
    3.2.5 Prewetting   52
    3.2.6 Application Rates   53
    3.2.7 Storage and Handling of Materials   53
    3.2.8 Environmental/Infrastructure Effects   57
    3.2.9 Chemically Impregnated Pavement   59
    3.2.10 Problems with Chemicals and How to Avoid Them   60
3.3 Mechanical Methods   60
    3.3.1 Distinction between Snow Removal and Ice Removal   61
    3.3.2 Equipment for Snow and Ice Removal   62
3.4 Thermal Methods   62
    3.4.1 What Are Thermal Methods?   63
    3.4.2 Design Factors   68
    3.4.3 Disposal of Snow; Snow Melters   70
3.5 Control of Drifting Snow   70
    3.5.1 Snow Fences   71
    3.5.2 Living Snow Fences   74
    3.5.3 Other Vegetative Barriers   76

Chapter 4. Maintaining a Safe, Trafficable Roadway   77

4.1 Traction   77
4.2 Importance of Tire/Pavement Friction   77
4.3 Definition of Friction; Coefficient of Friction   78
4.4 Factors Affecting Friction   79
4.5 Slip Ratio   80
4.6 Measurement of Friction   81
4.7 Measurement Methods   82
    4.6.1 Stopping Distance   82
    4.6.2 Deceleration   83
    4.6.3 Wheel Slip   83
4.7 Critical Values of Pavement Friction   84
4.8 How to Increase Friction Values   85
4.9 Abrasives for Friction Improvement   85
    4.9.1 Materials and Their Effectiveness   86
    4.9.2 Application of Material   88
    4.9.3 Effect of Abrasives on Air Quality   89

Contents   **vii**

         4.9.4   Othe Problems; Cleanup   89
         4.9.5   Storage of Materials   89
4.10   Mechanical Methods   90
4.11   Chemical Methods   90
         4.11.1   Anti-icing   91
         4.11.2   Deicing   93
4.12   Thermal Methods   93

Chapter 5. Snow and Ice Control Equipment   **95**

5.1   Introduction   95
5.2   Snow Removal Equipment   95
         5.2.1   Evolution of Equipment   96
         5.2.2   Plows   97
         5.2.3   Power Brooms   104
5.3   Ice Removal/Control Equipment   104
         5.3.1   Spreaders for Solid Materials   104
         5.3.2   Applicators (Spreaders) for Liquid Chemicals   109
         5.3.3   Ice-cutting Blades   110
         5.3.4   Underbody Blades   110
5.4   Snow and Ice Control Vehicles   110
         5.4.1   Mechanical Requirements   111
         5.4.2   Visibility   111
5.5   Other Equipment   112
         5.5.1   Graders   112
         5.5.2   Front-end Loaders   112
5.6   Sidewalk Snow Removal   112
5.7   Snow Loaders   112

Chapter 6. Weather and Pavement Condition Intelligence   **115**

6.1   It's a War Out There   115
6.2   Sources of Weather/Climate Information   115
         6.2.1   Government Sources   116
         6.2.2   Private Sources   116
         6.2.3   Agency Sources   117
6.3   Road Weather Information Systems   117
         6.3.1   What Is an RWIS?   117
         6.3.2   System Description   117
6.4   Radiometers   121
6.5   Thermography (Thermal Mapping)   122

Chapter 7. Decision Making for Snow and Ice Control (by Duane E. Amsler)   **123**

7.1   Introduction   123
7.2   Decision Criteria   123
         7.2.1   Air Temperature   124
         7.2.2   Pavement Temperature   124

　　　　　7.2.3　Characteristics of Snow and Ice Events　125
　　　　　7.2.4　Start Time of Events　125
　　　　　7.2.5　Event Intensity　125
　　　　　7.2.6　Character of Precipitation　126
　　　　　7.2.7　Event Duration　126
　　　　　7.2.8　Wind Speed/Direction　126
　　　　　7.2.9　Time of Day/Season　126
　　7.3　Pavement Friction　127
　　　　　7.3.1　Measuring Pavement Friction
　　　　　　　　 (*See also* Chap. 4)　127
　　　　　7.3.2　Estimating Pavement Friction　127
　　　　　7.3.3　Pavement Friction as a Decision-making
　　　　　　　　 Tool　128
　　7.4　Command and Communication Center　128
　　　　　7.4.1　Weather and Road Condition Data　128
　　　　　7.4.2　Use of Agency Resources　129
　　　　　7.4.3　Communication with the Public and
　　　　　　　　 Road Users　129
　　7.5　Command of Snow and Ice Control Operations　130
　　7.6　Reporting Requirements for Snow and Ice Control　131
　　　　　7.6.1　Pre-event Reporting　131
　　　　　7.6.2　During-the-Event Reporting　132
　　　　　7.6.3　Poststorm Reporting　132

Chapter 8. Managing Snow and Ice Control
　　　　　　(by Duane E. Amsler and Robert S. Pryzby)　**133**

　　8.1　Introduction　133
　　8.2　Agency Objectives　133
　　8.3　Management Control　134
　　8.4　Snow Policy and Operations Plan or Manual　134
　　8.5　Some Legalese for Keeping Out of Trouble　136
　　　　　8.5.1　Tort Liability　136
　　　　　8.5.2　Negligence　137
　　　　　8.5.3　Standard of Care　137
　　　　　8.5.4　Notice　138
　　8.6　Other Considerations in Forming a Policy　138
　　　　　8.6.1　Snow Ordinances　138
　　　　　8.6.2　Route Prioritization　139
　　　　　8.6.3　Road Closures　139
　　　　　8.6.4　Interagency Cooperation　140
　　　　　8.6.5　Police Agencies　140
　　　　　8.6.6　Other Agencies　141
　　　　　8.6.7　Emergency Management Office　141
　　8.7　Level of Service　142
　　8.8　Snow and Ice-Related Accidents　143
　　8.9　Training　144
　　8.10　Preseason Preparations　145
　　8.11　Off-Season Preparations　145
　　8.12　Use of Weather Information　146
　　8.13　Communication　147
　　8.14　Records　147

Contents

8.15 Operational Strategies   148
8.16 Snow Disposal in Cities and Towns   149
8.17 Performance Evaluation   149
8.18 Postscript   150

Chapter 9. Trucks for Snow and Ice Control
(by Richard W. Hunter)   **153**

9.1 Introduction   153
9.2 Prime Movers   154
9.3 Snowplow Truck Component Considerations   156
    9.3.1 Frames   156
    9.3.2 Axles   160
    9.3.3 Wheels and Tires   163
    9.3.4 Engines   163
    9.3.5 Transmissions and Driveline   165
    9.3.6 Brakes   166
    9.3.7 Cabs   167
    9.3.8 Lighting Systems   168
    9.3.9 Hydraulic Systems   169
    9.3.10 Bodies   169
9.4 Installation of Snow and Ice Equipment   170
    9.4.1 Hitches   170
    9.4.2 Wing Mounts   171
    9.4.3 Controlling Costs in Specifying Attachments   171
9.5 Rotary Plow Carriers   172
9.6 Reliability and Maintainability   173
9.7 Alternative Prime Movers   173
9.8 Conclusion   175
9.9 Specification Checklists   175

Chapter 10. Railroad Snow and Ice Control   **181**

10.1 Introduction   181
10.2 Current Snow Clearance and Snow Control Methods   184
    10.2.1 Main Line   184
    10.2.2 Yards and Terminals   185
    10.2.3 Switches   185
    10.2.4 Highway Grade Crossings   187
    10.2.5 Tunnels and Snow Sheds   187
10.3 Snow Removal Equipment   188
    10.3.1 On-track   188
    10.3.2 Off-track   189
10.4 Electric Systems   190

Chapter 11. Airport Snow and Ice Control   **191**

11.1 Introduction   191
11.2 Equipment   192

                11.2.1 Blade Plows   192
                11.2.2 Brooms   193
                11.2.3 Rotary Plows (Snowblowers)   193
                11.2.4 Spreaders   193
                11.2.5 Liquid Applicators   194
        11.3 Chemical Uses   194
                11.3.1 Aircraft Deicing/Anti-icing Fluids   195
                11.3.2 Chemical Control on Runways, Taxiways, and Ramps   195
                11.3.3 Liquid Chemical Application Rates   195
        11.4 Friction Testing and Reporting   197
        11.5 Snow Desks   199
        11.6 Reference Materials   199

Appendix A. Reference Materials   **201**

Appendix B. Chemicals and Their Properties   **209**

References   **263**

Glossary of Frequently Used Terms   **269**

Index   **279**

Introduction

What is the purpose of a book on snow and ice removal and control? What can anyone tell me about something as simple and low-tech as plowing snow and tossing some chemical on the road to burn off the ice? Isn't all snow the same? Furthermore, what can anyone do about all that white stuff that falls on us, or the ice that forms on our roads, airports, and railroads? That has been the attitude for the many years of brute force snow clearance and "tucking the salt" onto any ice, whether real or imagined, on traveled ways. There was some justification for that view until fairly recently because there were no innovative and imaginative alternatives to those brute force methods. Snow removal, or rather the lack of it in major storms, has figured in political disasters for some city officials, and more than one mayor of a major American city has suffered political repercussions as a consequence of the unacceptable performance of his or her (yes, both have been on the receiving end) snow removal response. One such incident led to the publication of an account of one city's unfortunate experience in 1969. It carried the appropriate title "The Political Properties of Crystalline H_2O: Planning for Snow Emergencies in New York" (Savas 1973).

Though there has been a low level of research into improved techniques and equipment for snow and ice control since the end of World War II in the United States, and to a greater extent in Europe and Scandinavia, it wasn't until the last decade of the twentieth century was well advanced that a concerted and coordinated effort to address this deficiency was initiated. Annual costs of snow and ice removal and control have been climbing for many years and now total well over $2 billion in the United States. This is money that produces no permanent improvement in a community, but the expenditures have been accepted philosophically because of the clearly adverse effect on people's lives and their commerce if nothing is done to maintain reasonably unimpeded travel in the winter. The increasing use by industry of just-in-time production methods has added to the demand for year-round unimpeded travel, and most shipments are by truck.

With costs becoming an increasing burden, government agencies both large and small have realized that their lagging technology must be improved to cope with the increasing demands for more efficient operations. Research has now resulted in better designs of equipment for the removal task. New technologies have introduced new practices which are now reducing the deleterious environmental consequences of the massive amounts of chemicals once used for deicing. Annual salt use for highway ice control had been holding at around 10,000,000 tons (9,072,000 t), but this has been increasing since 1993 because of the new demands made on the highway system and also because of climatic factors.

With the fruits of new research reflected in new and improved equipment and techniques, no book can be considered completely up to date. The technology of snow and ice control is not static. Therefore the emphasis in this book is on the fundamentals of snow and ice control, that is, the available technology and the scientific underpinnings of that technology, rather than a cookbook account of what actions to take or what procedures to follow. In our view, this will better prepare those most intimately involved with the critical task of ensuring the best performance of our transportation systems under adverse winter conditions. The goal is to provide the reader with a firm foundation that will enable him or her to assess the value of current and proposed equipment and techniques.

Units Used in This Book

The primary dimensional units used in this book are English units. Units conforming to SI (Système Internationale), commonly called the metric system, are given in parentheses following the English units. The two temperature scales, Celsius (often called centigrade) and Fahrenheit, are used interchangeably, but the equivalent temperature always follows in parentheses. Presented below are two aids to facilitate conversion between the two systems of temperature units.

Calculation Method

Derivation of the formula for converting either from degrees Fahrenheit to degrees Celsius or from degrees Celsius to degrees Fahrenheit can best be explained by use of a diagram (see Fig. 1).

Introduction

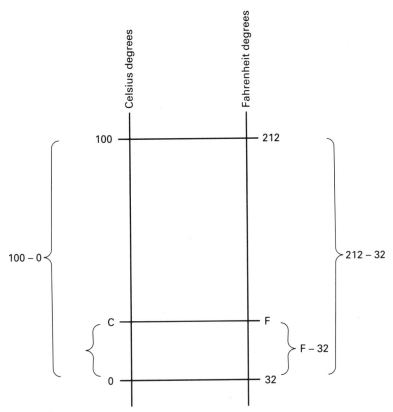

Figure 1.

The left vertical line represents Celsius (C) degrees; the right vertical line is Fahrenheit (F) degrees. The boiling point and freezing point of water provide convenient reference temperatures. We know that water boils at 100°C (212°F) at standard atmospheric pressure, and freezes at 0°C (32°F). These end points are shown on their respective scales. Finding any intermediate temperature between these extremes involves setting up a proportion. This proportion is in effect the ratio of two lengths, (1) the distance from the lower end points (the freezing temperatures) to the C↔F line, and (2) the distance between the two end points of freezing and boiling. This yields

$$\frac{F-32}{C-0} = \frac{212-32}{100-0} \qquad (1.1)$$

and simplifying gives

$$F = \frac{9}{5}C + 32 \qquad \text{or} \qquad F = 1.8C + 32$$

and

$$C = \frac{5}{9}(F - 32) \tag{1.2}$$

Using the decimal value 1.8 in the conversion from Celsius to Fahrenheit provides a handy anchor for making a quick mental conversion, since 1C° (Celsius degree) = 1.8F° (Fahrenheit degree), or 10C° = 18F°. Recalling that 0°C and 32°F refer to the same point on the temperature scale enables us to make the jump to 10°C equaling the sum of the starting point on the Fahrenheit scale, 32°, then adding the number of Fahrenheit degrees equivalent to 10 Celsius degrees, or 32 + (10)(1.8) = 32 + 18 = 50°F. Temperatures below freezing are handled in the same way, e.g., −15°C is 32 + (−15)(1.8) = 32 − 27 = 5°F. We can also use these convenient reference points for making additional mental estimates of equivalent temperatures. With practice, conversions can be made quickly and almost effortlessly.

Celsius temperature	Fahrenheit temperature
−40	−40
−20	−4
−10	14
−0	32
−10	50
−20	68

Appendix A contains a chart for converting temperatures either from Celsius to Fahrenheit or from Fahrenheit to Celsius and offers another method that can be used. It will be necessary to interpolate to find intermediate values when they are not listed.

Caution: When converting a temperature such as occurs in the statement "this temperature is only 60°F from the phase change," do not include the 32°F offset in the calculation, as you would if you unthinkingly applied Eq. (1.2). The conversion in this case is 60/1.8 [or (5/9)(60)] with the result 33.3°C, not (5/9)(F − 32) = 15.6°C. The distinction between the two cases is that one, the phase change example above, is a difference value. The other case, in which Eq. (1.2) is used, is for conversion of a fixed value.

Application rates of chemicals and antiskid materials are commonly expressed in terms of lane miles (la-mi) in the United States, e.g., 100 lb/la-mi. The standard width of a lane is 12 ft, and that is the measurement used for calculations of area applications; a lane mile is equal to 63,360 ft². Some agencies and publications use a value of 9 ft for a lane,

Introduction

since ice control materials generally will not have to cover the entire width to achieve an acceptable degree of traction improvement. It is common outside the United States to express application rates in terms of amount of material spread per unit area, and grams/square meter (g/m^2) is most frequently used; 1 g/m^2 = 13 lb/la-mi. An alternative metric unit, analogous to the North American lb/la-mi, is the kilogram/lane-kilometer, or kg/la-km. Other unit conversions are tabulated in Appendix A.

Acknowledgments

The unsung heroes of winter maintenance are the dedicated men and women who will leave the comfort of their beds or the warmth of the garage or office to make the roads passable and safe under the worst of weather conditions. Conversations with many have shown how much it means to them to perform these duties with the utmost skill. Much in this book has been distilled from the combined experience of many of these operators and their supervisors and managers.

Chapter 7 was written by Duane E. (Dewey) Amsler, formerly with the New York State Department of Transportation. Contributors to Chapter 8 were Dewey Amsler and Robert S. Pryzby, Prairie Village, Kansas. Chapter 9 was written by Richard W. Hunter, Illinois Department of Transportation. Their particular insights have served to add much substance to this book. Ultimate responsibility for incorporating these chapters into the book, and for the opinions, statements, and interpretations of data from many sources, falls on me.

The resources of the library of the Cold Regions Research and Engineering Laboratory were of immense value in tapping the extensive world literature on winter maintenance. The staff, in particular Nancy Liston and Elizabeth Smallidge, were of great assistance in calling my attention to many items, for which I am very grateful.

The snowfall maps of the United States in Chap. 2 have kindly been provided by Nolan J. Doesken of the Colorado Climate Center.

L. David Minsk
Hanover, NH

1

The Basics of Snow and Ice Control

A Quick Course in Methods, Procedures, and Facts to Help Do the Best Job of Coping with Snow and Ice

1.1 The 10 Principles of Snow Removal: Be a Bully— Hit It While It's Down. Match Technique to the Condition

What to consider doing:

Based on Snow Properties

1. *Know and understand the properties of snow that will make your job easier.* Snow has many unusual properties that will affect the magnitude of the effort required to clear a road or airfield. These include hardness, compaction, age hardening, water content, temperature instability, and the effects of mechanical agitation, among others. The type of snow (and ice) most prevalent in your region may influence the selection of equipment types (spreaders, plows, size of plow trucks), the treatment to be used to achieve the goal of maintaining a safe pavement-tire friction level, and the plowing techniques.

2. *Be aware of seasonal variations that affect the condition of your roads and any snow or ice on them.* Sensitivity to temperature is the most significant property of snow and has a tremendous influence on all its other unusual properties. Most solid materials we encounter every day have melting points far removed from the temperatures we experience, even on the hottest summer day. Most metals, for example, have melting points several thousand degrees above the temperatures we live in. Snow and ice, on the other hand, exist within a very few degrees of their melting point. The coldest temperature recorded in the continental United States is about $-70°F$ ($-57°C$), a mere 100 Fahrenheit (57 Celsius) degrees from the point where solid ice becomes liquid water. Most major population areas around the world do not experience temperatures much below $-40°$ (interesting fact: This is the same temperature on both the Fahrenheit and Celsius scales), and a very high percentage of snow removal takes place within a few degrees of the melting point of ice. Even in those regions with the lowest midwinter temperatures, the transition seasons of spring and fall bring many snowfall/freezing rain events near the freezing point. Just as water evaporates more rapidly on hot summer days than in the dead of winter, so the changes in a snow cover take place more rapidly at temperatures near the freezing point. Actually, the same property of water accounts for this behavior in both winter or summer: The rate at which water vapor migrates is a consequence of its vapor pressure, and this increases as the temperature increases. Water vapor migrates from the small ice crystals to the large, and in so doing increases the number of grain contacts. The result is an increase in the hardness of the snow.

3. *Reduce energy loss due to snow compaction during plowing by using a well-designed plow.* New snow traps a large volume of air within the spaces between snow grains. When the snow is compacted by mechanical loading, energy is absorbed by grain-to-grain fracturing and by compression of the air spaces. This energy loss can be as high as 40 percent during plowing. Displacement plows must accomplish three actions, each of which can result in energy loss: (1) shear the snow from the surface on which it rests (this may be shearing within the snowpack if snow remains on the road), (2) lift the snow above the level of the pack, and (3) change the snow flow from parallel to the truck plowing direction by approximately 90° to cast the snow to the side, away from the cleared path. Energy loss will be a minimum if the snow is disturbed as little as possible. This will occur to the greatest extent if the snow flow is sheetwise, i.e., if the layer of snow being removed is not agitated, but flows as a continuous sheet. Many plows have a moldboard that is close to vertical over the plowing face within

1 to 2 ft (30 to 60 cm) of the road surface, and these designs will result in high snow compaction. The most efficient design from the standpoint of energy loss minimization incorporates a cutting surface that enters the snowpack at an angle of around 30 to 45° from the horizontal. This design will lift the snow with minimum compaction. The next requirement for reducing snow compaction is a smooth transition from a forward direction to a direction to the side. One-way plows handle this transition most efficiently because of their conical shape, i.e., their moldboard increases in height toward the outboard end. In contrast, reversible plows have a cylindrical shape; they have the same geometry on both sides, and the directional transition results in turbulent mixing of the snow, with consequent energy loss.

4. *Don't let disturbed snow remain in an intermediate location; move it promptly to its final disposal area to avoid the consequences of age hardening.* Snow has many unusual properties, one of which is age hardening. New snow is usually low-density and soft, and can be compacted to a great extent. As a snowpack ages, a natural process of vapor migration that increases the number of contacts between the individual snow grains occurs within the pack. The result is an increase in density and hardness (hardness is the resistance to penetration). This process, called age hardening, occurs over a period of hours to weeks, depending on the air temperature. When low-density new snow is disturbed by mechanical action, such as wind drifting or plowing, the clock is speeded up and age hardening occurs in a matter of hours, if not minutes. This rapid increase in density and hardness produces snow that becomes nearly as difficult to penetrate as ice. Anyone who has walked on a windrow of snow freshly left by a passing plow has experienced this rapid increase in hardness; you remain on top of the snow, in contrast to sinking into nearby soft new snow. Thus, if multiple passes are required to clear an area, the effort will be reduced if the plowed snow is moved promptly to its final disposal location. There can be situations in which a windrow has been left in an intermediate location too long (perhaps less than an hour) and the plow truck which created it has insufficient power to cut into it. The best advice: Don't delay widening the cleared area.

5. *Clearing snow from roads should start with preventing it from getting there in the first place.* Time was when a snow fence was an invitation to extra work, expense, and frustration with the poor results. That was back when the standard fence, the 4-ft- (1.2-m-) high picket model, placed willy-nilly in the available right-of-way, either got buried, fell apart, or dropped the snow where it caused a bigger problem than it had before the fence was installed. Modern methods of controlling blowing snow make use of engineered barrier systems, placed intelli-

gently based on analysis of local climatology and terrain. Depending on prevailing wind conditions, seasonal snowfall, fetch (length of the open area serving as a blowing snow source), and available right-of-way, a fence using horizontal slats with a porosity (the open space compared to the total fence height) of about 50 percent and anywhere from 4 to 16 ft (1.2 to 4.9 m) high may be the solution. Removing obstacles that contribute to drifts across a road may be possible, since wind can be a friend as well as an enemy: It can scour a road of any snow that reaches it. Use of living snow fences may also be a cost-effective approach, though this may take several years to become effective. Temporary barriers in the form of corn stubble have been used successfully in some farm areas. When designing new roads or rebuilding existing ones, elevate the road above the surrounding grade and flatten the slopes to enlist the wind in scouring the surface.

Based on Ice Control Materials

6. *Know the characteristics of materials that are available to treat your road conditions.* Ice control materials range from freezing-point depressants that lower the temperature at which water freezes to temporary friction improvers such as abrasives. Important material properties of chemicals include composition, form, eutectic temperature/composition, lowest effective temperature, melting effectiveness, corrosion effects, surface effects, moisture absorption, toxicity (for humans and flora/fauna), environmental effects, storage and handling, cost, and availability.

- *Composition:* What is it made of, and does it have any undesirable properties?
- *Form:* Is it available as a solid or a liquid?
- *Eutectic temperature/composition:* At what concentration of the chemical and at what temperature does it freeze completely?
- *Lowest effective temperature:* When does the chemical cease to be useful as a freezing-point depressant? Does it remain useful for a reasonable length of time (say 1 h) that the agency has determined is necessary, given prevailing conditions?
- *Melting effectiveness:* How much ice is melted by the chemical?
- *Corrosion effects:* Will metals be corroded by the chemical, and if so, to what extent?
- *Surface effects:* Will the chemical affect friction because of a slippery film?
- *Moisture absorption:* Will the chemical absorb moisture from the air?

The Basics of Snow and Ice Control

- *Toxicity:* How toxic is the chemical to humans and to flora/fauna?
- *Environmental effects:* What other effects will the chemical have on the natural and built environment?
- *Storage and handling:* Does the material have to be stored under special conditions (e.g., protected from moisture), and is protective gear required during handling?
- *Cost and availability:* Are the benefits accruing from application of the material greater than its cost? Is it readily available in quantities sufficient for the agency's needs?

7. *Select the best material to apply for your varied conditions based on economics, environmental demands, climate, level of service, material availability, and application equipment availability.* The three most important criteria influencing the selection of a treatment to enable unimpeded traffic on a road are

- The importance of the road, i.e., the level of service, which will determine the extent of the resources to be devoted to it
- The environmental constraints, which will influence the choice of a chemical or nonchemical approach, and the chemical characteristics needed to satisfy any constraints
- The climatic characteristics of the area

Roads with low traffic densities in a low-snowfall or mild climatic region may be allowed to build up a short-lived snowpack and treated with abrasives when needed to improve friction. As traffic densities increase and the duration of periods below freezing increase, an anti-icing approach has been demonstrated to be very successful at either reducing the cost of treatment or reducing the duration of reduced friction. An anti-icing strategy is one that seeks to prevent the formation of a strong ice-pavement bond by timely application of an ice control chemical before or during the initial stage of precipitation. This contrasts with a deicing strategy, which is to allow snow to accumulate on a road to a specified depth prior to any treatment such as plowing or application of abrasives, then use chemicals to break up the snowpack to clear to bare pavement.

Based on Road Characteristics

8. *Know the surface state of your roads during the winter season, the most important aspect of which is the surface temperature.* Weather reports and forecasts provide much meteorological information, including the air temperature. Though this is useful as an indicator of temperature trend, it is not as valuable to the maintenance manager as is the pave-

ment surface temperature. Pavement temperature will invariably lag any change in air temperature by anywhere from several minutes to several hours, depending on the heat content of the road structure, the road's exposure to sun or shade or dark of night, to some extent the wind speed, and whether the surface is wet or dry. It is the pavement temperature that will largely determine the type of treatment required. If the pavement is cold—that is, less than 20°F (−7°C)—and dry, and cold, dry snow is falling, the best strategy is usually to let wind and traffic action disperse any slight accumulation, then plow deeper accumulations without any further treatment. Bonding of cold, dry snow to cold pavement is usually very weak, and plowing will be sufficient for adequate removal. As the pavement temperature approaches 32°F (0°C), the likelihood of snow bonding to the pavement increases, and plowing along with the application of a chemical may be the best strategy. That decision should be made with as much information as possible, and that means knowing the pavement surface temperature. Several methods for measuring pavement surface temperature are now available to maintenance personnel: detectors installed in the pavement surface course which transmit the temperature to a remote station, portable handheld infrared radiometers, and vehicular-mounted infrared radiometers. Embedded detectors are fixed in place and provide point measurements, whereas the handheld and mobile radiometers can be moved from point to point. Each has advantages: Embedded detectors are valuable for their early warning capability, as well as for their continuous monitoring. The infrared radiometers are useful for surveying the temperature characteristics of road segments, a process called thermal mapping. They can also provide the manager with the information necessary to select the best treatment for the existing conditions.

9. *Investigate installation of weather sensors in and alongside the roadway to provide real-time information about the microclimate; document the road system characteristics that will have a significant effect on snow and ice control—for example, pavement type, solar exposure, "cold sinks," snow drifting potential, traffic characteristics, bridges and overpasses, and safety hazards such as railroad crossings.* Snow and ice control requirements will vary widely over a road network. Hilly terrain may provide hollows where snow accumulates to a greater extent than elsewhere, pavement surface temperatures will differ because of exposure and the varying thermal properties of the road structure, and proximity to water bodies may increase the probability of ice formation. Keeping track of the changing microclimates of the road network is very important. As mentioned in paragraph 8 above, knowing the pavement surface temperature is very important for making an informed decision

regarding the best treatment strategy. Embedded ice detectors can indicate the presence of snow or ice and measure the surface temperature and are valuable as warning devices. However, they provide only point measurements, and because of cost constraints relatively few may be installed over a road network. The thermal characteristics of the pavement between embedded ice detectors may vary over a wide range, and consequently the optimum snow and ice control strategy may vary. It is useful to map the characteristics of the road network to indicate potential problem locations. Since pavement temperature is the single most important factor influencing an ice control strategy, measuring the temperature of the pavement between detectors will provide the manager with better information for decision making. Infrared radiometers offer a rapid noncontact means of temperature measurement. Scanning the pavement temperature over a network and plotting the values, a practice called thermal mapping, enables quick identification of warm and cold spots and thus provides a method for selecting locations for installation of embedded pavement ice detectors and temperature sensors. More detail can be found in Chap. 6.

Based on Weather

10. *Investigate use of meteorological services to provide essential information for scheduling preventive and corrective responses to winter storm events.* Technology now available can provide real-time information about the microclimate of road segments to meteorologists, who can then make specific forecasts that will be much more reliable than national or regional forecasts.

1.2 The Six Principles of Ice Control: When and How To Use Chemicals; How Much to Use

1. *Anti-icing is generally more efficient and effective than deicing.* Deicing is required when a snowpack that is strongly bonded to the pavement develops or ice is permitted to form on pavement that has received no bond-prevention treatment. Breaking the strong bond by cutting or scraping is difficult, and much of the pack or ice will remain. Very thin layers of bonded snow or ice can be removed completely with a small amount of chemical, but thicker layers [1/4 in (6 mm) or more] require much greater quantities of chemical to reach the

ice-pavement interface. Deicing therefore generally leads to excessive use of chemicals. Application of a bonding inhibitor to the pavement prior to snowpack or ice formation will eliminate the need for deicing over a wide range of conditions. This is the essence of anti-icing, a technique that has been used in the past to some extent but has become more practicable over the last few years with the availability of more precise local weather information. Good information is necessary to time the chemical application before precipitation starts, or very soon after it does. However, the extra attention required for an anti-icing strategy may not be justified on low-volume roads.

2. *Liquid ice control chemicals offer advantages over solid chemicals in many cases.* Anti-icing is most effective when most or all of a lane is treated uniformly. Spraying a liquid permits good control of application rate and coverage and will result in much less material loss during application and from traffic compared to solid chemical. A liquid will remain an effective bond inhibitor for hours or days under many conditions, depending on precipitation and traffic. It has prevented early morning flash icing on bridges in central California for several days after application.

3. *Chemicals are more effective than abrasives in most situations.* Abrasives serve only to provide a temporary increase in friction between a snow- or ice-covered pavement and rubber tires. They will not eliminate the hazard resulting from the precipitation. Observations made during a Federal Highway Administration test program showed clearly that abrasives applied on well-traveled roads, whether used with or without chemical additives, did not improve the friction on a road receiving properly timed anti-icing treatments. In fact, there were many instances in which friction was reduced by the abrasives.

4. *Prewetting solid chemical before application speeds action and reduces loss.* If deicing becomes necessary, solid chemical will generally perform better and faster in breaking up pack. Chemicals lower the freezing point of ice only when in liquid form, and wetting the solid before it is placed on the road will speed up the process of its becoming liquid. The liquid film on solid chemical particles also helps reduce material loss during application by serving as a glue that prevents material from bouncing off the traveled way and therefore is helpful even when solid chemical is used as an anti-icing treatment.

5. *Temperature, most importantly that of the road surface, affects chemical performance.* Temperature is important in two respects: (1) It will take longer for a solid chemical to dissolve and lower the freezing point as the temperature drops, and (2) the quantity of chemical needed to melt ice increases as the temperature drops. Ultimately a temper-

ature will be reached at which no amount of chemical will cause melting (the "eutectic point" of the chemical). Every chemical has a unique maximum concentration of the solution and minimum melting temperature. For salt, the eutectic temperature is $-6°F$ ($-21.1°C$) and the eutectic composition is 23 weight percent of the solution.

    6. *All chemicals have some deleterious effect on the environment; they should be applied in moderation.* Some chemicals will react chemically with portland cement concrete (pcc)—ammonium compounds are among those that were used for ice control before this became apparent. The chemicals most likely to be used for ice control—the chlorides (calcium, magnesium, potassium, sodium), the acetates (sodium, potassium, calcium, magnesium), the formates (sodium, potassium), and urea—will not appreciably affect pcc that has been made with an air-entraining agent and has aged for a year before chemicals are applied. Unprotected steel reinforcing bars (rebars) may be subject to corrosion by the chloride chemicals if the salt solutions reach them. This will be minimized by the use of epoxy-coated rebars, with at least a 2-in (6-cm) thickness of concrete above the rebars. However, all chemicals that lower the freezing point of water will promote deterioration of pcc pavements by physical action. The creation of a chemical concentration gradient within the concrete will lead to internal expansion forces, which in time may cause spalling or cracking. Sealing the concrete to prevent intrusion of the ice control chemicals will reduce the deterioration. Asphalt concrete is not affected to the same extent as pcc, although any exposed aggregate may suffer some degradation with time. Reducing the quantity of chemicals applied to the pavement will reduce the concentration as dilution occurs with precipitation and thereby lessen the potential for deterioration. All chemicals will have some effect on the roadside environment also. Sodium affects soil structure adversely, whereas calcium, magnesium, and potassium will benefit the soil. Water- and snow-laden chemicals splashed on vegetation cause harm.

2
The Nature of Winter Precipitation

A Matter of Materials Handling

2.1 Characteristics of Snow and Ice

Water, both as a liquid and in its solid forms of ice and snow, is the most ubiquitous compound on the surface of the earth. The oceans occupy 70 percent of the earth's surface and represent a little over 97 percent of all the water on earth; water bound as ice on the polar ice caps represents another 2 percent. Though a simple molecule of two atoms of hydrogen and one atom of oxygen, water has unusual properties that still baffle scientists. These properties influence life itself, since the cells of living creatures are composed of up to 80 percent water. Of more immediate concern, knowledge of the material properties of snow and ice will reduce the work involved in their handling and removal. The eight critical properties of snow, and the critical property of ice at ordinary temperatures, are discussed in Sec. 2.5.

2.2 Scope of the Task of Snow and Ice Removal

Snow and ice have a huge effect on the economy of all countries where winters last several months. In the United States, an estimated $2 billion is spent every year by highway agencies to maintain the nation's roads and streets in a safe, trafficable condition during the winter. A single major northern airport may spend more than $2 million over the course of a winter season to keep runways, taxiways, and ramps reasonably clear of snow and ice. Additional costs are incurred by motorists and shippers who experience travel delays and by businesses that are forced to close temporarily because of winter storms of sufficient magnitude to paralyze a community or region. Estimates of such costs are notoriously inaccurate, as are estimates of the environmental costs. Better estimates of the direct cost of infrastructure damage resulting from the use of chemicals for snow and ice control have been made; they range between $2,100,000,000 and $4,400,000,000 annually (TRB 1991, p. 63). Vehicle corrosion accounts for the largest portion of the infrastructure costs (90 percent); damage to bridge decks and other structures such as parking garages accounts for the remainder.

The impact of snow and ice is highly variable; it is affected by population density, the topography of a region, the road network, and, of course, the climatic factors of snowfall rate, temperature regime, hours of daylight, and wind. The historical record of snowfall in the United States can show to some degree how significantly winter weather affects life, and the maps in Fig. 2-1 depict this graphically. Not to be overlooked are the expectations of the driving public. All developed countries depend on a complex transportation network—whether rail, air, or road—for their economic well-being, and snow and ice must be prevented from interfering with the functioning of this network to the fullest extent possible.

2.3 Snow and Ice Removal as a Materials Handling Problem

What do we mean by snow and ice "control"? We cannot control snow and ice short of raising the temperature of the atmosphere above freezing in winter, or heating all the surfaces we wish to remain snow-free, or constructing protective roofs over our cities and highways. All these mechanisms actually do play a role, oddly enough: The "heat island" effect of major metropolitan areas modifies the climate so that snowfall is reduced, or rain (or freezing rain) falls rather than snow. In New England and the Midwest of the United States (and in other countries), covered bridges are a time-honored method for reducing snow accumulation on the roadway. Snow sheds are used even today on railroads and highways in mountainous areas of Europe, Japan, and

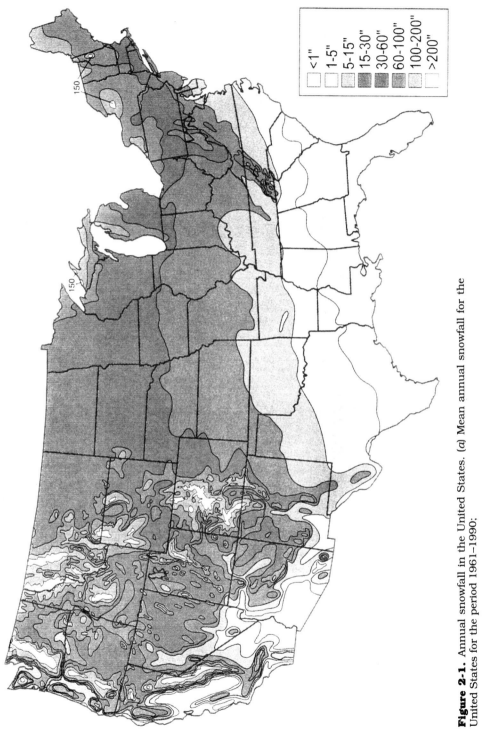

Figure 2-1. Annual snowfall in the United States. (a) Mean annual snowfall for the United States for the period 1961–1990;

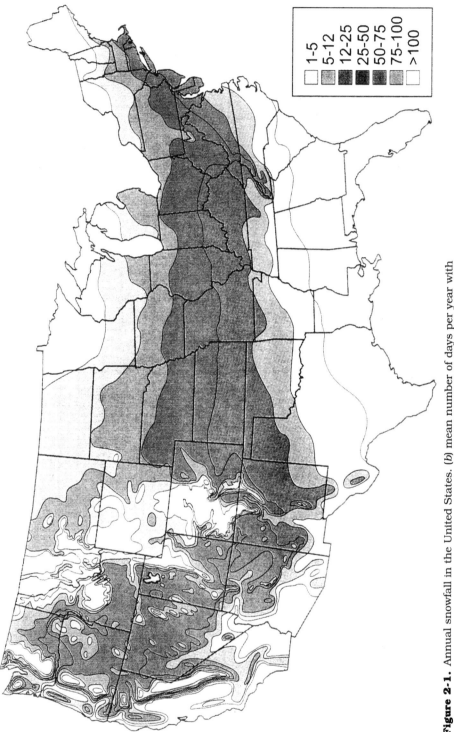

Figure 2-1. Annual snowfall in the United States. (*b*) mean number of days per year with 1 in or more of snow on the ground for the period 1961–1990;

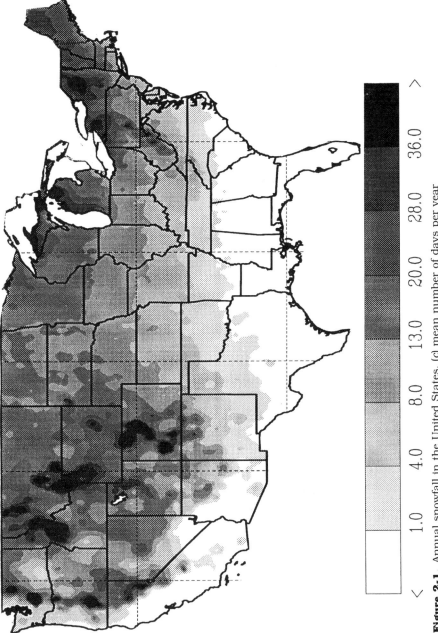

Figure 2-1. Annual snowfall in the United States. (c) mean number of days per year with snowfall of 1 in or greater based on data for the 1961–1990 period.

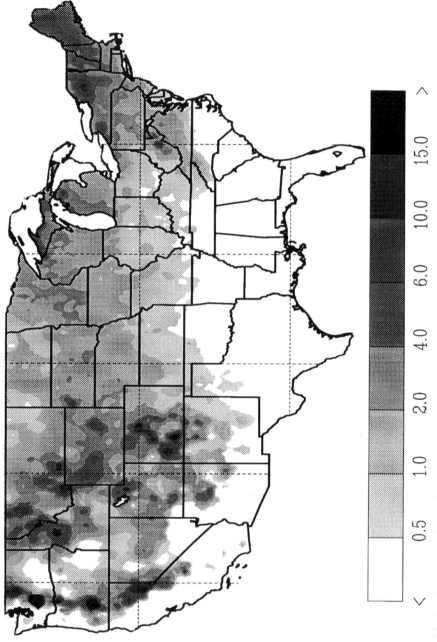

Figure 2-1. Annual snowfall in the United States. (*d*) mean number of days per year with snowfall of 5 in or greater based on data for the 1961–1990 period (Doesken and Judson 1996).

North America for the same purpose, and for preventing snow avalanches from blocking these transportation arteries. That is one meaning we can give to control of snow and ice: modifying the environment in some deliberate manner to reduce the deleterious effects that excessive accumulations will have on the functioning of a system. That system can be a navigable waterway where shipping is constrained by ice accumulation, an electric power plant where the trashbars of the cooling water intake become obstructed by ice, or the inlet of an air ventilating system for a structure in a cold climate which accumulates sufficient snow or ice to choke off air flow. Solutions to those problems are covered in other manuals. In this book we will consider only transportation systems: highways, airfields, and railways.

The control of snow and ice on transportation facilities is largely a task of moving huge volumes and masses of material. Thus we can think of snow and ice removal as a materials handling problem—indeed, some have termed snow a solid waste. Whatever you call it, clearance of the immense amounts accumulating on transportation facilities is necessary to meet today's dependence on rapid and unhindered year-round movement of people and shipment of goods. The quantities are truly staggering: 6 in (15 cm) of snow on a 12-ft- (3.66-m-) wide lane 1 mi (1.61 km) long weighs between about 200,000 and 400,000 lb, or 100 to 200 tons (90,720 to 181,440 kg). A city with 500 lane-miles (la-mi) (805 lane-kilometers) of roads would have to contend with 50,000 to 100,000 tons of that seemingly fluffy white powder in just one 6-in (15-cm) snowfall. Contrast this with the total amount of rock and earth excavated for the construction of the Panama Canal: about 400 million tons over a 33-year period (McCullough 1978). The properties of snow and ice that directly affect the methods of their control are described in the next section, on the premise that the more you know about your adversary, the better you are prepared to beat it.

2.4 The Material Snow

Both snow and ice are forms of the solid state of water. Snow is a porous, permeable aggregate of ice grains that can be predominantly single crystals or close groupings of several crystals. Snow and ice differ considerably in their properties, and so they are described separately. Ice is covered in detail in Sec. 2.6.2. However, they have in common the unusual fact that they exist very close to their phase change, i.e., close to the temperature at which the solid becomes liquid. In the popular mind, all snow is the same. A closer examination will reveal many subtle differences in a snow cover, depending on temperature, water content, degree of compaction, impurities, and age. The Inuit

(Alaskan and Canadian Eskimos) recognize these subtleties and have dozens of words to refer to specific types and conditions of snow.

2.4.1 Birth of a Snow Crystal

Snow originates in subfreezing clouds. Air containing water vapor, transported from large bodies of water or from soil and plants, expands and cools as it rises. The amount of water vapor the air can hold decreases as the temperature drops until saturation is reached, i.e., the relative humidity (RH) is 100 percent. The water vapor then condenses on "condensation nuclei" consisting of salt particles from the ocean and droplets of sulfuric and nitric acids from natural and industrial sources. As the air cools below 32°F (0°C), the droplets become supercooled, since they are too small to freeze at that temperature. Concurrently, some water vapor changes directly to the solid form and deposits ice on so-called freezing nuclei consisting of clays blown skyward by wind erosion. As the cloud cools further, it becomes supersaturated (the RH is greater than 100 percent). The ice crystals grow at the expense of the water droplets because of the latter's greater vapor pressure. As the crystals grow, they begin to assume the familiar hexagonal shape of snowflakes as a result of the molecular structure of the ice crystal. But the form of the crystal as growth continues depends on the temperature and the degree of supersaturation (Doesken and Judson 1996). There are many variations, as shown in Fig. 2-2. When the crystals become massive enough, they begin to fall at rates varying from 0.8 to 8.5 ft/s (0.25 to 2.6 m/s) (Nakaya 1954). For comparison, raindrops fall at a rate of about 23 ft/s (7 m/s) (Wallace and Hobbs 1977).

The lowest temperature recorded in the lower United States, −70°F (−57°C), at Rogers Pass, Montana, is only 102°F (57°C) from the solid-liquid phase change. Seldom do temperatures in more populated areas in North America drop below −40°F (−40°C), only 72°F (40°C) from the phase change. This difference is even less during the frequent periods when the temperature is within a few degrees of the freezing point. This is a remarkable distinction from the common solid metallic materials surrounding us, most of which change phase at temperatures well over 1000° above commonly experienced temperatures (see Table 2-1). Most of the physical properties of snow and ice are strongly influenced by temperature, and thermodynamic instability is an inherent characteristic of these materials. This will become apparent in the discussion of their physical properties. These can be grouped broadly into mechanical, thermal, and electrical. Only the first two types are relevant to snow and ice control and will be described.

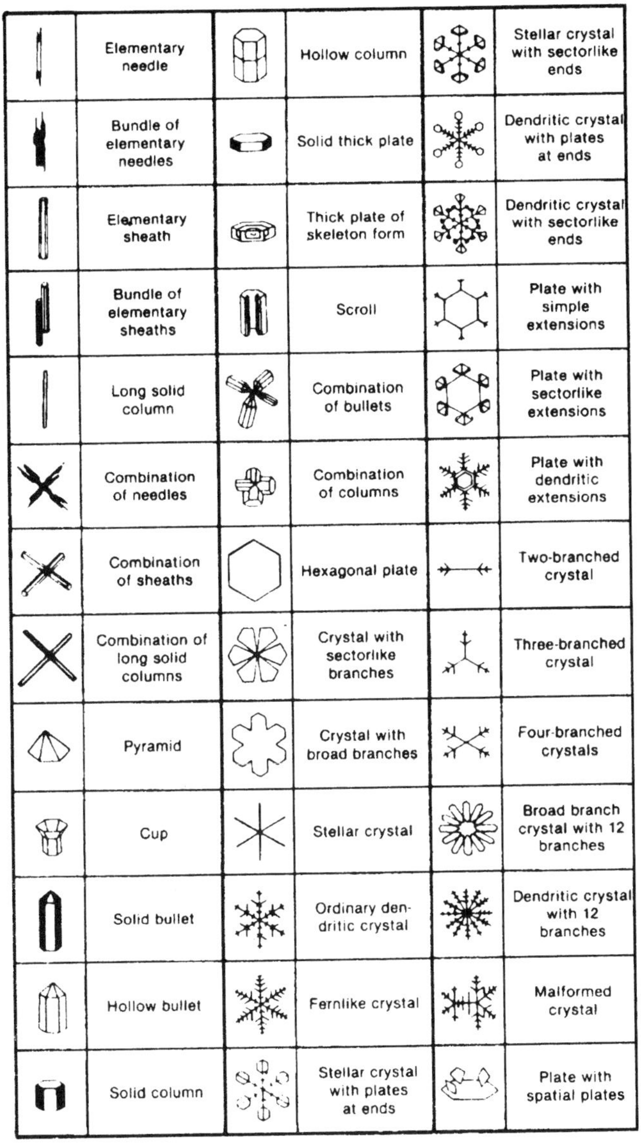

The Nature of Winter Precipitation

Elementary needle	Hollow column	Stellar crystal with sectorlike ends
Bundle of elementary needles	Solid thick plate	Dendritic crystal with plates at ends
Elementary sheath	Thick plate of skeleton form	Dendritic crystal with sectorlike ends
Bundle of elementary sheaths	Scroll	Plate with simple extensions
Long solid column	Combination of bullets	Plate with sectorlike extensions
Combination of needles	Combination of columns	Plate with dendritic extensions
Combination of sheaths	Hexagonal plate	Two-branched crystal
Combination of long solid columns	Crystal with sectorlike branches	Three-branched crystal
Pyramid	Crystal with broad branches	Four-branched crystals
Cup	Stellar crystal	Broad branch crystal with 12 branches
Solid bullet	Ordinary dendritic crystal	Dendritic crystal with 12 branches
Hollow bullet	Fernlike crystal	Malformed crystal
Solid column	Stellar crystal with plates at ends	Plate with spatial plates

Figure 2-2. Drawings of the forms of snow and frozen particles.

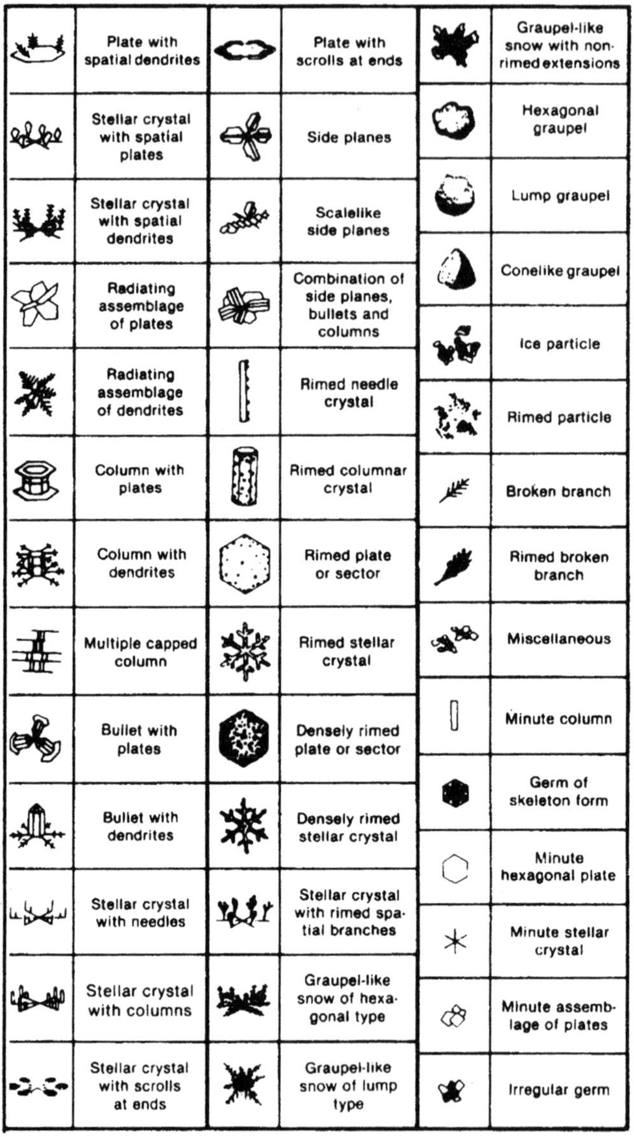

	Plate with spatial dendrites		Plate with scrolls at ends		Graupel-like snow with non-rimed extensions
	Stellar crystal with spatial plates		Side planes		Hexagonal graupel
	Stellar crystal with spatial dendrites		Scalelike side planes		Lump graupel
	Radiating assemblage of plates		Combination of side planes, bullets and columns		Conelike graupel
	Radiating assemblage of dendrites		Rimed needle crystal		Ice particle
	Column with plates		Rimed columnar crystal		Rimed particle
	Column with dendrites		Rimed plate or sector		Broken branch
	Multiple capped column		Rimed stellar crystal		Rimed broken branch
	Bullet with plates		Densely rimed plate or sector		Miscellaneous
	Bullet with dendrites		Densely rimed stellar crystal		Minute column
	Stellar crystal with needles		Stellar crystal with rimed spatial branches		Germ of skeleton form
	Stellar crystal with columns		Graupel-like snow of hexagonal type		Minute hexagonal plate
	Stellar crystal with scrolls at ends		Graupel-like snow of lump type		Minute stellar crystal
					Minute assemblage of plates
					Irregular germ

Figure 2-2. (*Continued*) Drawings of the forms of snow and frozen particles. (*Koh 1989*)

Table 2-1. Phase Change (Solid to Liquid) Temperatures for Some Metals and Ice

Material	Melting point (°F)	Melting point (°C)
Aluminum	1220	660
Steel, low carbon	2760	1516
Copper	1980	1080
Lead	621	327
Ice	32	0

2.5 The Important Properties of Snow

The snow properties which must be considered in devising a comprehensive snow removal strategy are tabulated here; the ranking is arbitrary, since each property is dominant at some time during the life history of a snow cover.

- Density
- Hardness (or strength)
- Compressibility
- Cohesiveness
- Adhesiveness
- Temperature instability
- Age hardening
- Effects of mechanical agitation

2.5.1 Density

This is the mass per unit volume, a measure of the amount of material in a given volume. Conventionally, the density D is

$$D = M/V$$

where M = weight (mass) of snow, lb (kg)
 V = volume of snow, ft³ (m³)

A fixed quantity of snow will undergo a change in density as its volume changes. One should keep in mind the inverse relationship between volume and density: As volume changes, the density changes in the opposite direction. That is, when the volume of the constant

Table 2-2. Densities of Different Types of Snow

Type of snow	Density (kg/m³)	Density (lb/ft³)
New, fresh	50–60	3–4
Old	70–100	4–6
Compacted	160	10
Becomes ice	830	52

mass decreases, the density will increase—the same amount of snow occupies a smaller volume.

Values of density for snow range from very low, 3 lb/ft³ (50 kg/m³), for fresh new snow to about 37 lb/ft³ (600 kg/m³) for snow that has been disturbed by some mechanical action, such as plowing or wind. Old snow that has not been compacted by vehicle tires or other mechanical loading will not generally exceed a density of 25 lb/ft³ (400 kg/m³). When the density of a snowpack reaches 52 lb/ft³ (830 kg/m³), the air passages in the normally porous snow become discontinuous and the mass becomes impermeable. Snow scientists have adopted the convention of calling this the point at which snow changes to ice. Table 2-2 summarizes the ranges of densities for various kinds of snow.

2.5.2 Hardness (Strength)

Hardness of snow is related to its resistance to structural collapse or to penetration. The stresses causing structural collapse can result from tensile, compressive, or shear forces. In all these failure mechanisms, the bonds between adjacent snow crystals break, resulting in a sudden loss of *cohesion*. Tensile forces pull the snow crystals apart, compressive forces crush them, and shear forces cause failure by a sliding action. Compressive failure sometimes causes a snowpack to crumble. Mechanical snow removal stresses result mostly from compressive and shear forces.

Snow hardness depends on the grain structure and temperature. Grain structure, in turn, depends on the snow density and the extent of bonding between adjacent grains. Snow is nearly cohesionless when it first falls, i.e., individual grains have few points of contact, and where they do touch, the contact area is very small. At this stage, hardness (strength) is very low. Over time, bonds develop and grow at the grain contact points. This involves a migration of water molecules from one point in the structure to another, and the rate of the migration process is highly dependent on the temperature of the surroundings. The migration proceeds rapidly as the temperature approaches the melting point, slowly at low temperatures [below about 10°F (-12°C)]. As the temperature rises toward the melting point, liquid water begins to coat the individual snow grains. The density may remain the same or increase slightly if the volume decreases, but strength will decrease much more rapidly because of lubrication between the grains. Conversely, as temperature drops, hardness increases because of the stronger intergranular contacts, although it will take some time for the hardness to increase because of the slower molecular activity.

The harder the snow is, the more difficult it is to penetrate with a blade plow or to disaggregate with a rotary plow. Hardness, as measured by unconfined compressive strengths, ranges from less than 1 lb/in^2 (6.9 kPa) for new snow with a density of 6 lb/ft^3 (100 kg/m^3) to 30 lb/in^2 (207 kPa) for well-bonded snow with a density of 25 lb/ft^3 (400 kg/m^3). Snow hardness can be determined by measuring the resistance to penetration of a probe. Snow scientists use several devices for this purpose. One, the Canadian Hardness Gage, consists of a spring which can be compressed to provide a range of forces for pushing a circular disk into the snow. Several combinations of disk diameter and spring compression levels can be chosen to cover the wide range of snow-cover hardnesses. A good estimation of hardness can be obtained by the methods shown in Fig. 2-3, using your hand or pocket items. However, there is a very good correlation between compressive strength and density for cold snow (i.e., snow with no free water filling the pore spaces). Therefore, measurement of the density may suffice as an indication of hardness under those conditions. Freeze-thaw cycling of a snowpack may result in the formation of ice lenses, i.e., layers of ice formed when meltwater trickles down to a depth where it refreezes. This can increase the strength of a snow cover significantly in a short period of time.

2.5.3 Compressibility

A common mechanism of densification of snow, with the resulting increase in hardness, arises from the second important property of snow, its compressibility. The fact that a volume of snow can be reduced as much as eight times by compaction significantly affects the snow removal effort. In contrast, soils can be compacted only to a volume half the original. Since the compressive strength of new low-density snow is very low, this snow will readily compact under loading. The granular structure will collapse in stages until a density of 31

Testing snow hardness The hardness class is based on which of the following can easily be pushed into the snow.	Hardness class	Corresponding hardness gauge reading (g/cm²)
Closed fist covered with glove	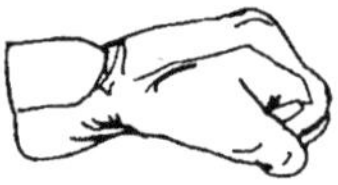 Very soft	0–500
Flat extended hand with glove	 Soft	
Extended glove-covered index finger	 Medium hard	500–2500
Pencil	 Hard	2500–5500
Knife	 Very hard	>5500

Figure 2-3. Quick determination of snow hardness can be made using the hand or pocket tools (Koh 1989).

lb/ft³ (500 kg/m³) is reached. At this point the snow grains are closely packed, and stresses then increase rapidly because any further compaction occurs not by the breaking of intergranular bonds but by deformation of an increasing number of snow grains. Once snow becomes ice at the density mentioned in the last section, 52 lb/ft³ (830 kg/m³), it will not ordinarily compact further, since all excess air has been expelled. Energy is absorbed by friction between snow grains in a collapsing snow structure, and by acceleration of the particles and their fracture into smaller grains. This is a factor in snow removal, since plowing with most present-day blade plows cannot be accomplished without some compaction of the snow. At plowing speeds above about 20 mi/h (32 km/h), compaction can absorb as much as 34 percent of the energy produced by the plow truck engine (Pell 1994, p. 9).

The pores of cold, dry snow are filled with air and water vapor. In wet snow, the ice grains are also coated with liquid water in thermal equilibrium. The amount of water held in the snowpack as either bound or free water profoundly affects the work of removal and the response of the snow to traffic action. Under load, the larger particles grow at the expense of the smaller ones because of temperature differences induced by pressure melting at the points of intergranular contacts. Compaction can be quite rapid until temperature equilibrium is established. At that point, further compaction occurs by packing and crystal deformation. Impurities in the pore space between particles affect the compressibility of the snow. Salt, or any ice control chemical, applied to a snow cover will act in this way. Salt will reduce the rate of compaction; the higher the concentration of salt in the pore water, the less the snow will compact (Colbeck 1978).

The compressibility of snow is a factor in the selection of snow removal equipment. Both displacement (blade) plows and rotary plows must be mounted on carriers that have adequate power to push them into the snowpack, lift the snow, and cast it to the side. Considerable work is required to compact snow. When a blade plow pushes into a snow layer on a road, a plastic (pressure) wave propagates into the snow ahead of the plow. Energy is dissipated by the compaction that results. Research has shown that the work of compaction increases as the speed of compaction increases, and that this increase is nonlinear, i.e., as speed increases in equal increments, the work required to do this goes up faster. A computer analysis demonstrates this in the graph shown in Fig. 2-4 (Hanson 1991).

2.5.4 Slush

It has been found experimentally that the amount of free or unbound water (i.e., the water in the pore spaces) that will be retained in a wet snow cover at a temperature of 32°F (0°C) after it has been allowed to drain is about 6 percent. If the snow lies on an impermeable surface,

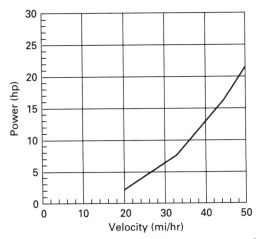

Figure 2-4. Power absorbed by compaction of snow increases as plowing speed increases.

such as a hard-surfaced road, water may be retained beyond what would be present in the freely drained condition. The resulting mixture is called slush. The particles are surrounded by a continuous water film, and some particles may even be suspended in the liquid. Slush has no compressive strength and will readily flow. Snow with a free-water content of less than 15 percent is readily compacted by traffic and forms a slippery surface; with a free-water content between 15 and 30 percent, snow does not compact but remains in a soft, loose state on the road surface. At around 30 percent, snow adheres to tires. When the free-water content exceeds 30 percent, snow is readily removed from the pavement by splashing (Schaerer 1970). At around this upper level of water content, the snow may even float on the water.

2.5.5 Cohesiveness

You can't make a good snowball with fresh dry snow because it doesn't stick together, it crumbles easily. The snow grains lack cohesion, and so the snowball is cohesionless, because there are relatively few intergranular bonds. (The term *cohesion* refers to the attraction between similar materials; *adhesion*, to be described in the next section, refers to attraction between different materials.) Similarly, when cold, dry snow is initially deposited, it lacks cohesion because of the lack of intergranular contacts. As snow ages, the number of grain contacts increases, and thus the cohesion increases. Plowing with either a blade plow or a rotary plow breaks up much of the original snow mass. The compressive strength of the displaced snow increases because of the cohesion of the increased number of closely packed particles. Disaggregated snow particles produced by a rotary plow, in fact, are so cohesive that they bond together on contact. This can plug a discharge chute with a bulky snow mass if the opening isn't sufficiently large. Who hasn't experienced walking on new snow and sinking deeply into it? If the same snow is disaggregated by plowing, or even by wind action, it rapidly becomes so hard because of cohesion that you can walk on top without sinking in. Age hardening is one process in which cohesion plays a role (see Sec. 2.5.8).

2.5.6 Adhesiveness

Snow will stick to many surfaces until the temperature approaches the melting point. The number of grain contacts on the supporting surface influences the extent of the attachment, as does the temperature of the interfacial boundary. New cold, dry snow falling on a cold, dry pavement will not adhere, or bond, strongly. Wind and vehicle turbulence will frequently clear light accumulations from a road surface, for example. As the temperature rises above about 20°F (-6.7°C), the amount of liquid water surrounding the snow grains increases. This increases the adhesion of the snow to its support surface because of the increase in the contact area and the attraction the water has for both contact surfaces. Many types of plastics are so-called low-energy surfaces; Teflon (polytetrafluoroethylene) is representative of this class of fluorocarbon plastics. Snow and ice do not adhere strongly to such surfaces because of their low wettability; thus the points of contact are very small.

2.5.7 Temperature Instability

The proximity of the melting point of snow to its temperature in nature, in contrast to the wide separation of those two temperatures in most other materials we are familiar with, as previously described, is a major reason for the great variation of snow properties with temperature. It affects the amount of liquid water coating the particles, which in turn influences compressibility, hardness, density, and cohesiveness. *This characteristic of snow is probably the most important single factor affecting its properties, not the least of which is the work required for removal or control.* See Sec. 2.5.10 for additional information on the thermal properties of snow.

2.5.8 Age Hardening

Another consequence of the temperature instability of snow is its change of structure with time. Random thermal oscillation of the molecules at the ice surface induces migration of water vapor in the pore spaces, with the molecules moving from the small particles to the larger. With time, therefore, an undisturbed snow mass will densify (its volume decreases) and increase in hardness and compressive strength up to some limit as the crystals grow to a maximum size of about 3 mm. Bond strength between crystals then drops rapidly, leading to formation of corn snow, or rotten snow. This snow has very low cohesion and is easily penetrated.

Another type of snow that may form in undisturbed snow is depth hoar. This is the result of vapor migration within a snowpack when there is a strong positive thermal gradient [greater than 18 Fahrenheit (10 Celsius) degrees]. A positive gradient is one in which the bottom layer is warmer than the top. A very distinctive snow crystal forms, with grain sizes generally between about $\frac{1}{16}$ and $\frac{5}{16}$ in (2 and 8 mm) but sometimes as large as $^{11}/_{16}$ in (15 mm). Bonding between particles is very poor, and consequently the strength of this type of crystal is low. A depth hoar layer represents a weak shear plane and is a major cause of avalanches. It is also very difficult to drive either wheeled or tracked vehicles over a snowpack with a depth hoar layer.

2.5.9 Effects of Mechanical Agitation

The densification/hardening process that occurs in a natural snow layer is accelerated by the disaggregation, or breaking up, of the layer by mechanical agitation. This can occur through two processes: wind action, causing blowing and drifting snow, and plowing with either displacement (blade) or rotary plows. Snow deposited by one of those processes will rapidly increase in strength because of the sudden increase in the number of particle contacts and the higher packing density. In effect, this is accelerated age hardening. You may have experienced the result of this process by not breaking through the surface when you walked on a recently formed snowdrift or on a snow windrow recently cast by a plow. This is of practical importance in snow removal; the work of clearance will be reduced if the snow is plowed to its final resting place rather than being rehandled at a later time, when removal will take extra work because of the rapid hardening.

2.5.10 Thermal Properties of Snow

Heat Capacity. When heat is added to a body, its temperature will rise. The heat capacity is a measure of this; it is the ratio of the quantity of heat added to an object to the increase in its temperature. It is expressed as British thermal units per degree Fahrenheit (Btu/°F) [joules per degree Celsius (J/°C)]. Materials differ from one another in the quantity of heat required to produce a given temperature rise in a given amount of material. Dividing the heat capacity by the quantity of material (its mass) gives a value that is specific to the material; it is called the specific heat capacity, or sometimes just specific heat. Units are Btu/lb-°F (J/kg-°C). Snow is a good thermal insulator as a result of the large amount of air trapped within the mass. Because air has low specific heat, the heat capacity of snow is due almost entirely to the ice. Though it varies somewhat with temperature and purity, a value usable for engineering purposes is 0.503 Btu/lb°F or 2106 J/kg-°C. The slight influence of temperature is demonstrated by the value of heat capacity at -22°F (-30°C): 0.45 Btu/lb•°F or 1884 J/kg-°C. For comparison, the value for water is nearly 1 Btu/lb-°F, or 4187 J/kg-°C. Another expression for the heat content of a substance is specific heat ratio, defined as the ratio of the specific heat capacity of the substance to that of water. The specific heat ratio is numerically equivalent to specific heat capacity. As a ratio of numbers with the same units, it is dimensionless.

Thermal Conductivity. This property denotes the rate at which thermal energy will be transferred through a substance. It is expressed as the quantity of heat energy which will be conducted through a unit area and unit thickness in unit time with a thermal gradient of one degree. The units are Btu•in/ft²•h•°F (W/m•°C). Thermal conductivity in snow is much more complex than in ice because of the fewer contact points between ice grains and the influence of air, which serves to insulate the particles. It is heavily dependent on the structure and texture of the snowpack and the air flow rate through it. Heat is transferred in snow by conduction through the interconnected ice grains; by conduction, convection, and radiation across the air spaces; and by migration of water vapor by sublimation and condensation. Satisfactory values for conductivity can be calculated for low-density snow, i.e., snow having a density no greater than 21 lb/ft³ (340 kg/m³) (Bader, 1962):

$$k = 0.0068r^2 \text{ cal/cm•s•°C} \quad \text{or} \quad k = 2.845r^2 \text{ W/m•°C}$$

$$k = 0.00508w^2 \text{ Btu•in/ft}^2\text{•h•°F}$$

where k = thermal conductivity
r = density, g/cm³
w = density, lb/ft³

Some values of thermal conductivity at a few densities are given in Table 2-3.

Because of the air trapped in the crystals, snow is a good thermal insulator; its thermal conductivity is low. The thermal conductivity of wet snow (snow close to the melting point with all spaces between ice grains occupied by liquid water) is zero because it has the same temperature throughout the mass; it is isothermal. Heat transfer in this case can occur by only two processes: transmission of radiant energy or percolation of meltwater which will release its latent heat upon

Table 2-3. Thermal Conductivity of Low-Density Snows

	Density, kg/m³ (lb/ft³)				
	50 (3)	100 (6)	150 (9)	200 (12)	300 (19)
k (cal/cm•s•°C)	1.70×10^{-5}	6.8×10^{-5}	1.53×10^{-4}	2.72×10^{-4}	6.12×10^{-4}
k (W/m•K)	0.0071	0.028	0.064	0.11	0.26
k (Btu•in/h•ft²•°F)	0.050	0.20	0.44	0.80	1.8

refreezing. As a consequence, wet snow is an efficient one-way valve: Heat can get in but not out. The importance of the thermal properties of snow and ice will be evident in Chap. 3 when thermal methods of snow and ice removal are described (Sec. 3.4).

2.5.11 Snow Is a Unique Material

The difference between earth materials (soil and rock, the other bulk materials handled in immense quantities) and snow is striking. As described earlier, the strength, or hardness, of cold snow correlates well with density. The strength of dry frozen ground, on the other hand, differs little from that of thawed ground. Only when soil contains water does its strength increase with freezing. The more the water (ice) content increases, the more the strength of soil approaches that of compacted snow or ice. Earth excavations are generally made in thawed ground, so it is appropriate to compare that material with snow. The contrast in properties displayed in Table 2-4 demonstrates the differences that directly and greatly affect the snow removal task. All the properties of snow detailed here make the work of removal or clearance greater than that for soil.

2.6 Water and Its Forms

2.6.1 Water

Water, the stuff of life on earth, exists in three forms, or states: the solid state (snow and ice), the liquid state (water), and the gaseous state (vapor or steam). They all have the same molecular composition: Two atoms of hydrogen and one atom of oxygen combine to form the molecule H_2O. These molecules are very tiny; one teaspoon of water (1 cm^3 or 0.06 in^3) contains as many molecules as the Atlantic Ocean contains

Table 2-4. Significant Properties of Snow that Distinguish It from Earth

Property	Thawed earth	Snow
Density	High (70–110 lb/ft^3)	Low–high, 3–50 lb/ft^3
Hardness	Low	Low–high
Age hardening	None	High
Compaction	Low	High
Adhesion	Low	High
Cohesion	Low–high	Low–high

teaspoonsful of water. Water has a number of properties which make it unusual and, in fact, explain its importance as the "stuff of life." Life originated in a liquid medium, and the salt (sodium chloride) concentration in human cells reflects this origin. The unusually high values of some of the measurable properties of water, such as its dielectric constant, specific heat, melting and freezing temperatures, and maximum liquid density at 39°F (4°C), are a reflection of a strong attraction between the hydrogen atoms which scientists call "hydrogen bonding." The details of this critically important type of bonding need not concern us, but their consequences are important and warrant description.

The water molecule looks like a slightly distorted dumbbell with an included angle of 105° (Fig. 2-5). Ice (Fig. 2-6) is an ordered array of water molecules fixed in a "puckered layer" structure, with each oxygen atom located at the center of a tetrahedron (a four-sided object like a pyramid) formed by hydrogen atoms. The openness of the structure accounts for ice having a lower density than water. If one views the ice molecule by looking down from "above" the layers, the characteristic hexagonal structure of snow crystals (Fig. 2-7) can be seen.

Differences in the energy levels of the molecules account for the three states. Changes in energy are necessary for changes of state, or phase transitions, to take place. Adding heat energy to ice increases the agitation of the molecules to the point where the rigid crystalline ice lattice breaks up into clusters of H_2O molecules. The reverse transition, from liquid to solid, releases the same amount of energy. The making of ice cubes in the refrigerator illustrates this. The freezing compartment of the refrigerator must be cooled below the freezing temperature, and modern appliances use the energy from electricity or gas to do this. Then when an ice cube is dropped in your drink, its melting requires heat energy, which it extracts from the drink, thereby

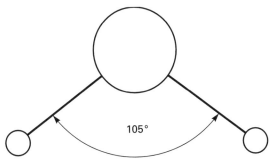

Figure 2-5. Water molecule; the large atom is oxygen (O), the two small atoms are hydrogen (H).

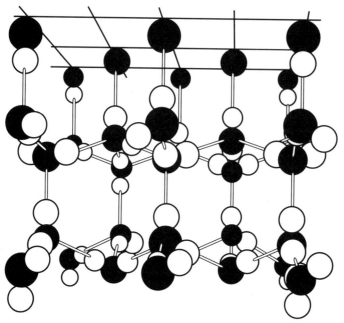

Figure 2-6. Ice at temperatures commonly found on earth has a structure in which the water molecules are arranged in a repeating pattern forming a crystalline solid with hexagonal symmetry. The oxygen atoms (black circles) are arranged as tetrahedra, one atom at the top with the three others forming the base of the tetrahedron. One hydrogen atom (white circle) is associated with each oxygen-to-oxygen link. This produces a "puckered layer" arrangement of the atoms and leads to a variation in the strength properties of ice depending on whether a force is applied parallel or perpendicular to the layered structure. This view is along the so-called basal plane of the ice. When viewed looking down on the layers, along what is called the optical axis, the characteristic hexagonal symmetry of the ice crystal is evident. This is shown in Fig. 2-7. [From Kepler, Johannes (1966) *The Six-Cornered Snowflake*. London: Oxford University Press. By permission of Oxford University Press.]

cooling it. The phase transitions for all three forms of water-substance have special names, and the energy levels all differ. These are listed in Table 2-5.

2.6.2 Ice and Its Properties

Ice on highways, runways, and structures poses a greater problem than snow because of the tenacity with which it adheres to surfaces that must be kept clear. As a consequence, ice is difficult to remove by

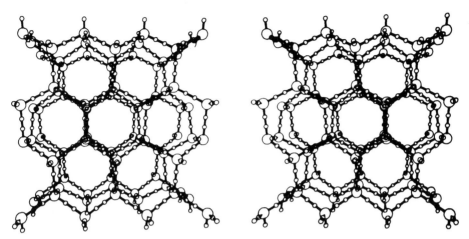

Figure 2-7. Characteristic hexagonal structure of snow as seen at the molecular level. This is a stereo pair for 3-D viewing. (*Hamilton and Ibers 1968*).

Table 2-5. Water-Substance Phase Transitions

Phase change	Name of change	Thermodynamic value		
		Btu/lb	kJ/kg	cal/g
Solid⇔liquid	Heat of fusion	144	334	80
(Liquid⇒solid)	(Heat of solidification or crystallization)			
Liquid⇔vapor	Heat of vaporization	1078	2500	598
(Vapor⇒liquid)	(Heat of condensation)			
Solid⇔vapor	Heat of sublimation	1222	2834	678
(Vapor⇒solid)	Heat of deposition			

mechanical methods, and alternative methods of removal or control, such as chemicals or heat, may increase the cost to as much as 10 to 15 times the cost of snow removal. Mechanical methods are frequently resorted to because of the necessity to avoid chemical contamination for some reason, because the appropriate chemical may not be available, or by choice, but these methods are also more costly for ice removal than for snow removal because they are slower and because of the extra wear on equipment and pavement. Mechanical scraping seldom clears the full width of the blade completely to bare pavement because of irregularities in the pavement and the built-in crown most roads have to improve drainage.

Physical Properties. Ice isn't a simple material. It can exist in 10 different forms. (For a tale of what would happen if a hypothetical form of ice he termed Ice IX existed, read Kurt Vonnegut's *Cat's Cradle*; life would never be the same.) Most exist at high pressures and very low temperatures. Only what is called Ice 1h, the hexagonal form, occurs at temperatures we experience in all temperate regions of earth. It would be necessary to drop below –112°F (–80°C) to find the form known as cubic ice. What follows, therefore, only applies to Ice 1h. When water freezes to form ice, it displays a very unusual property—it expands. This is in contrast to most materials, which contract upon freezing. This makes ice slightly less dense than water; therefore it will float. It is this anomalous behavior which enables aquatic life to survive the winter in lakes and rivers in cold climates. If these water bodies froze to the bottom, much of this life would be killed. Instead, ice forms on the surface and serves to insulate the lower depths, greatly reducing the thickness of ice growth.

Under ideal conditions in the laboratory, the density of ice in its purest form, with no foreign material (including air bubbles), has been determined to be 57.2 lb/ft^3 (917 kg/m^3). Pure water weighs 62.4 lb/ft^3 (1000 kg/m^3). Ice found naturally, whether arising from freezing rain or by freezing of ponded water, has air inclusions and perhaps other impurities that lower its density. As described in Sec. 2.5.1, the convention adopted by snow scientists is that when snow is compacted to a density of 52 lb/ft^3 (830 kg/m^3), it becomes ice because it will no longer allow the free passage of air through the mass—it becomes impermeable.

Occurrence. There are four common mechanisms of ice formation on pavements:

- Freezing rain
- Freeze-thaw of compacted snow
- Radiation cooling
- "Icings"—freezing of ponded water

Freezing Rain. If the pavement surface is at or below 32°F (0°C), rain falling on it may freeze if several conditions occur. Glare or glaze ice, a homogeneous, generally clear ice coating, is favored by a slow rate of freezing, large raindrop size, high rainfall rate, and only a slight degree of supercooling. This will occur because the raindrops flow enough to merge with one another and cover the pavement before freezing. This type of ice is one of the hardest to remove because of its intimate contact with the surface. Glaze ice generally forms at temperatures of 27 to 32°F (–3 to 0°C), although there have been cases reported where it formed at air temperatures as low as −5°F (−21°C) and as high as 37°F (2.8°C). Evaporation of water may aid in the freezing process; as indicated in Table 2-5, energy is required for the liquid to change to vapor (the heat of vaporization). A simple calculation shows that the evaporation of one volume of water falling at 23°F (−5°C) on a surface at 32°F (0°C) will freeze a volume 126 times larger. On the other hand, the liberation of heat which occurs on freezing may delay the freezing of cold rain on roads when temperatures are close to the freezing point.

Freeze-Thaw of Compacted Snow. Snow compacted by the passage of tires does not bond strongly to the pavement at low temperatures. However, if a thin layer of snow in contact with the pavement melts and then freezes, the ice layer that results will bond the compacted snow to the pavement. Frequently this layer is as strongly bonded to the surface as glaze ice, and is equally difficult to remove by mechanical methods.

Pressure Melting of Ice. Water expands when it freezes. Conversely, ice contracts when it melts. These physical properties of water substance led J. J. Thomson in 1849 to predict, on the basis of thermodynamic calculations, that the melting point of ice would be lowered by the application of pressure. This has been verified experimentally, and it has been found that the melting point will be lowered by 1 Fahrenheit degree under a pressure of 1090 lb/in^2 (or by 1 Celsius degree under a pressure of 13.5 MPa). This lowering is insufficient to cause any noticeable melting under tire loads. As an example, a 4000-lb (1815-kg) car exerts a pressure of only about 33 lb/in^2 (226 kPa) on the pavement, assuming a single tire footprint of 30 in^2 (194 cm^2), which would lower the freezing point under each tire by 0.003°F (0.002°C).

Some Terminology. Before describing the last mechanism of ice formation, radiation cooling, it will be necessary to explain some terms describing atmospheric processes. The *saturation point* of a moist air mass describes its capacity to hold an amount of water vapor that is in equilibrium at a specific temperature and pressure with a plane surface of either water or ice; *in equilibrium* means that there is no net gain in water molecules either on the surface by condensation or into the air mass by evaporation. An air mass can be *supersaturated,* which means that it will hold an amount of water vapor in excess of its saturation value. This is a very unstable condition and will not last long, particularly in the layer of air near the ground. The *dew point* is the temperature to which a moist air mass must be cooled for it to be saturated with respect to water. *Dew* is the water that is condensed when the dew point is reached.

Water may remain in the vapor state even though its temperature is below 32°F (0°C). The water droplets then are *supercooled* (this is the conventional term, though it is something of a misnomer since *super* means "above" or "greater," which suggests that a more appropriate term would be "undercooled"). The supercooled state is an unstable condition, and several situations can trigger instant solidification, including contact with an impurity which serves as a nucleus for the growth of an ice crystal, mechanical disturbance such as vibration, or so-called homogeneous nucleation, though this is not likely to occur unless the temperature is very low (in the vicinity of $-40°$). Pavements are invariably dirty, so water will not remain supercooled after contact.

Radiation Cooling. Just as water always flows downhill, a warm body gives up its heat to a cold body. Stand in a warm room near an uninsulated window and you will feel cold because your warm body (it better be!) is radiating its heat to a cold surface. A similar situation occurs on roads. The night sky is nearly a perfect blackbody (actually the temperature of outer space is about 3 degrees above absolute zero or 3 K, where K is kelvin, the SI unit of temperature) and very willingly will soak up heat radiated to it from earth. This doesn't have much effect when the air temperature is high enough to keep the road surface a few degrees above 32°F (0°C). When it does approach that point, however, the outgoing radiation may exceed the rate at which heat stored in the pavement and subgrade can maintain the temperature above freezing, and the road surface temperature will drop below the freezing point of water. This can happen even though the air temperature is several degrees above the freezing point of water.

The layer of air just above the pavement can hold a considerable amount of water vapor. The thin layer in contact with the pavement, however, can cool below the dew point. The water molecules that were moving so rapidly that they were able to remain suspended in the vapor state then slow down and drop onto the pavement. If the pavement temperature is above 32°F (0°C), dew will form. If it is below that temperature, some form of ice will be deposited. The rate and quantity of the deposition will depend on the amount of moisture in the air, the droplet sizes, and the rate at which the heats of condensation and fusion are given up. These variables will determine whether a road is transformed from a safe to a potentially hazardous condition.

Very small droplets, quite likely to be supercooled, will freeze nearly instantaneously upon contacting a cold pavement and form frost, or hoarfrost. ("Hoar" is derived from an old English word *har* or *hor*, meaning "old, gray.") Frost is a light, feathery deposit that is generally not tightly bonded to the pavement, since the contact area is very small. As a consequence, it may easily be removed by traffic action and pose no hazard. At the worst only a small amount of chemical will be needed to melt the deposit.

If, however, the water droplets are somewhat larger and hold more heat, they may condense on the pavement and spread out and join together in a thin continuous film before freezing, giving rise to black ice. This form of road icing is such a thin layer that it is transparent and therefore nearly invisible under most lighting conditions—hence its name. Two conditions are required for its occurrence: (1) a road surface temperature below freezing, and (2) a source of moisture in a calm air layer just above the road. These conditions are not uncommon in environments near bodies of water or other sources of moisture (sometimes supplied by industrial discharges). Or, after a few days and nights of clear, cold weather, a mass of warm, moisture-laden air moves in. As the air in the boundary layer immediately above the road cools, the vapor condenses in small droplets which are deposited on the road surface. If deposition is very slow, which is fostered by the stillness of the air, the droplets may become supercooled and freeze immediately upon contacting the road. This is the hoarfrost described in the previous paragraph. If, however, the droplets are larger and not significantly supercooled, or if deposition is more rapid and the droplets have time to flow and join together on the surface, a uniform coating of black ice may form. This type of ice is less common in a continental climate, since warmer, high water vapor content air occurs less frequently. However, there is the possibility of black ice forming from a light drizzle or very light falling rain.

During the night, frost will form on vegetation alongside a road before it forms on the pavement because foliage cools more quickly. There is often a very brief period just after dawn when the surrounding vegetation warms quickly but the pavement warms more slowly and may remain below freezing. The water vapor released by the vegetation may condense on the pavement, particularly on bridge decks or pavements of lighter construction because of their high thermal conductivity. These conditions favor the formation of frost or black ice in a very short time, giving rise to "flash icing" (Williamson 1969).

Rime Ice. Hoarfrost is sometimes confused with rime ice, but they arise in very different circumstances. Frost forms under very calm conditions, with the droplets falling onto a horizontal or nearly horizontal, surface under the force of gravity. Its appearance is feathery and generally rather thin. Rime, on the other hand, is a type of atmospheric ice formed by supercooled water droplets blowing onto a cold surface that is massive enough to dissipate the heat of fusion without rising above the melting point. The droplets freeze almost instantaneously and retain their nearly spherical shape. The accumulation will build out into the wind, and its density will increase with the wind speed, though the amount of water vapor in the air, the droplet sizes, and the air temperature will also have some influence. Thus this type of ice is found on the windward face of upright objects, such as signs. Whether there was wind during the formation of the ice deposit is the key to what type of ice

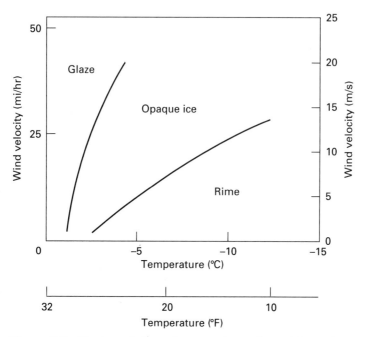

Figure 2-8. The type of atmospheric ice that is formed depends on wind speed and temperature.

results: Hoarfrost occurs in the absence of wind, whereas rime ice requires a wind. In addition to rime, glaze and a milky, opaque form of ice can form as the result of wind blowing water droplets against a surface. Figure 2-8 illustrates the relationship of wind speed and air temperature to the type of atmospheric ice that will form (Kuroiwa 1958).

Ice Adhesion. One of the problems caused by ice is that it usually really sticks to surfaces on which it forms. In fact, on metals and pavements the strength of adhesion may be greater than the bulk strength (the tensile strength) of the ice. That means that a failure or fracture will occur within the bulk ice rather than at the interface, with the result that some ice will remain on the metal or pavement. This is one of the reasons why mechanically scraping ice with a blade will not generally result in a bare surface, even if the surface is level rather than crowned as most roads are. It is not unusual for a portland cement concrete pavement to fail within the concrete rather than at the ice-pavement interface because of the greater strength of the ice bond. The bond strength is strongly dependent on the temperature of the ice-pavement interface. It is understandably weak in the vicinity of the melting point, but rapidly increases

The Nature of Winter Precipitation

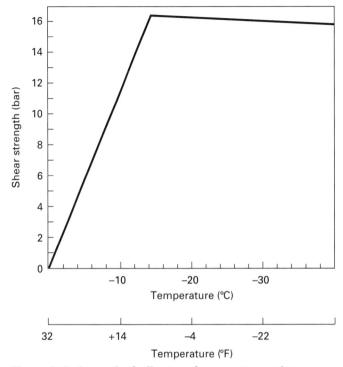

Figure 2-9. Strength of adhesion of ice variation with temperature. At −13°C (8.6°F) failure occurs within the ice (a cohesive failure). (*Jellinek 1972*)

as the temperature drops. For example, the strength of adhesion at 14°F (−10°C) is nearly twice as great as at 23°F (−5°C). This is shown in Fig. 2-9 (Jellinek 1972). Though this graph is for ice frozen on stainless steel, the effect on pavement materials would be similar, though the values would differ.

3
Snow and Ice Control Methods

3.1 Introduction

Broadly speaking, all snow and ice control techniques are classified as either chemical, mechanical, or thermal. Chemical methods include the application of a freezing-point depressant on a surface and incorporation of the freezing-point depressant within the surface itself. Mechanical methods include plowing or scraping and use of high-velocity air for scouring a snow-covered surface. Thermal methods involve applying heat to the surface from either above or below to remove snow and ice or to prevent its formation.

3.2 Chemical Methods

Chemicals, principally salt or sodium chloride,* have been used in the United States for snow and ice control on pavements for over 100 years. A letter to the editor appeared in the January 5, 1859 *New York Times* protesting, "What right have the stage proprietors to sprinkle salt the whole length of Broadway in order to remove the snow, and thus spoil the sleighing for the use of private citizens?" The editor responded that the "complaint must be taken *cum grano salis*....The truth is that salt is most efficacious...for the removal of freshly fallen snow."

---

*The term *salt* will be limited to sodium chloride in this book. Other chemicals that also meet the chemist's definition of a salt as the product of a strong base and a strong acid, such as calcium chloride and magnesium chloride, will be referred to specifically by name.

A column in *Scientific American* (Feb. 19, 1887) described the use of salt for snow removal in Paris. "The salt should be scattered on the streets as soon as the snow begins to fall fast," it advises. The recommended application rate was "in the proportion of about one drachm per square foot for each four-tenths of an inch thickness of snow fallen, or a larger amount if the temperature is low." Translated into modern terms, the equivalent is about 620 lb per lane-mile per inch (19 g/m² per centimeter) of snow. The intent was to melt all the snow to avoid plowing.

Initially little use was made of the application of straight chemical except in cities. Sand, cinders, and other abrasives were used to a much greater extent, and salt was added only to keep them from freezing so that they remained free-flowing. In 1941, however, New Hampshire commenced the use of straight sodium chloride as a general policy (Johnson 1946). A total of 3865 tons (3,506,000 kg or 3506 t) of salt was used the winter of 1941–42, though this figure includes an unspecified quantity mixed with sand. In 1971, the quantity used on a national basis had grown to approximately 9 million tons (8.2 million t) per year. The salt use curve (Fig. 3-1) started to climb steeply in 1962, but already by 1959 the environmental effects of deicing chemicals on roadside trees were beginning to appear, reportedly first in Minnesota (Sucoff 1975). Many reports have appeared during the last several years describing the effects of deicing chemicals on soils, vegetation, and structural materials. Environmental problems arising from the use of ice control chemicals are reviewed in Sec. 3.2.8.

3.2.1 Use of Chemicals for Snow and Ice Control

Chemicals are applied to pavements for three purposes: (1) to melt ice that has formed on pavement (deicing), (2) to prevent formation of ice (anti-icing), and (3) to prevent the buildup of "pack" (snow compacted by traffic action that becomes nearly as tightly bonded to pavement as ice, and that is frequently much thicker and more irregular). Both ice and compacted snow have coefficients of friction as low as 0.1 (see Sec. 4.3 for the definition of the coefficient of friction). Though friction-improving materials such as sand, slag, or cinders can be spread on a slippery surface, the problems that arise have discouraged their extensive use in many places; use of abrasives and their drawbacks are covered in more detail in Sec. 4.9. As a consequence, the greatest reliance has been on the use of chemicals. Several chemicals are available, but sodium chloride, because of its low cost, ready availability, ease of application, high solubility in water, and effectiveness as a melting agent at temperatures near 32°F (0°C), is most commonly used. Those advantages have contributed to the increased use of salt, which in

Snow and Ice Control Methods

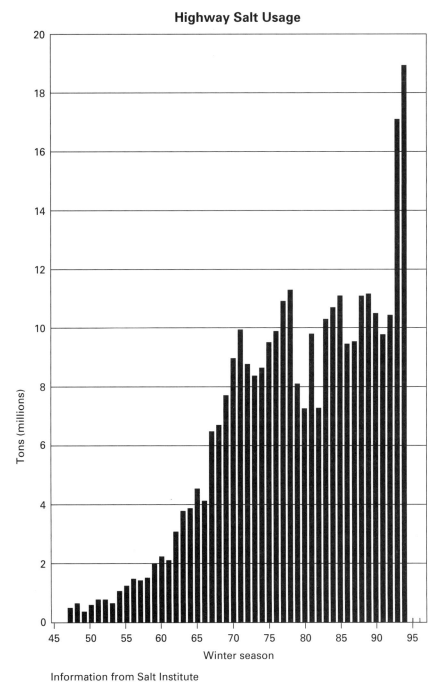

Information from Salt Institute

Figure 3-1. Quantity of salt applied to U.S. highways, 1975–1994.

many areas has led to elevated levels of sodium in drinking water and environmental stresses affecting fish and roadside vegetation, and contributes to corrosion of vehicles and structures.

3.2.2 How Chemicals Melt Ice

Melting is simply the transition from the solid state to the liquid. The solid state of water—ice—is in a lower-energy state than the liquid—water. Therefore, energy must be added to ice to transform it to the higher-energy state of water. This energy breaks the rigid crystalline structure of ice into the more disorganized free-flowing liquid. The most common source of energy for melting is thermal energy, that is, heat. But chemicals that are soluble in water have chemical energy, or chemical potential, that can produce melting. Some solids, like salt, absorb heat when dissolving—they are endothermic. Others, like calcium chloride, give off heat when dissolving—they are exothermic. Both types will melt ice because they depress the freezing point when a quantifiable amount of heat, the heat of fusion, is added to the system.

Imagine two trays of bees, one in the sun and one in the shade; some bees will leave the sunny tray but return, and some will fly to the shady tray. Bees will also be leaving the shady tray, but since most bees don't like to stay in a hot sun for long, more will return to it than will fly to the sunny tray, so the shady tray attracts more and more bees—eventually all of them. This is just what occurs when water freezes. Water molecules are continually pulling away from a water surface and flying into the air. Technically speaking, water has a higher vapor pressure than the moisture in the air. Some water molecules return to the surface and rejoin it, and an equilibrium is established, giving rise to a fixed vapor pressure dependent primarily on the temperature. This process is diagrammed in Fig. 3-2. Notice that the vapor pressure for water drops as the temperature drops because the molecular activity is less, just as the buzzing of the bees would decline. At temperatures below T_f, the normal freezing point of water, the vapor pressure of water continues to decrease along path AB. However, under ordinary conditions, ice begins to form at T_f, but it has a lower vapor pressure than water and is represented by path AC. This means that more molecules will collect on the ice surface than on the water, and the ice will grow at the expense of the water phase—more bees are attracted. Similarly, all the water will turn to ice when the heat stored in the water (the heat of fusion) is dissipated. So we see that water molecules act like bees, though not because they want to get out of the hot sun.

Addition of salt to water lowers its vapor pressure, as shown by the salt solution curve in Fig. 3-2. A simple explanation of why this occurs is that the molecules of the "impurity"—the sodium and chloride ions in

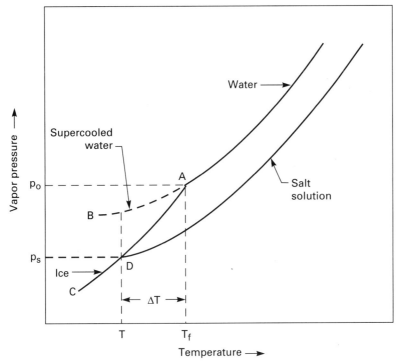

Figure 3-2. Depression of the freezing point.

this case—replace some of the water molecules at the surface, reducing the number available to enter the space above. At T_f the vapor pressure is below that of ice, so the ice will go into solution—in a sense, it flows downhill to the lower vapor pressure of the salt solution. Not until point D is reached will the salt solution curve intersect the ice curve, and because at this point the vapor pressures are the same, ice can start to form. ΔT is a measure of the freezing-point depression. A chemical that causes this is called a freezing-point depressant. The conditions that must be met for a chemical to depress the freezing point are (1) it must be soluble, i.e., it must dissolve in water; and (2) there must be sufficient energy to drive this process of solution. Substances that dissolve in water will depress the freezing point by a known, measurable amount for the same weight of substance added to the same amount of water. Table 3-1 illustrates this; it lists the freezing-point depression for a number of common materials. However, because some substances are less soluble than others, adding more material will not depress the freezing point by the same amount for all substances.

Another type of diagram, Fig. 3-3, describes the solubility of sodium chloride and relates the chemical concentration to its freezing tempera-

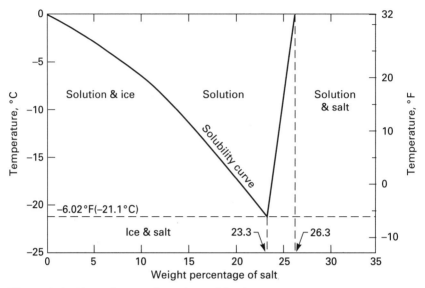

Figure 3-3. Phase diagram for sodium chloride.

ture. This is a solubility diagram, but it is also commonly known as a *phase diagram*. It is very useful for explaining the action of a soluble substance such as salt in depressing the freezing point. The horizontal scale at the bottom states the solution composition. The numbers represent the percentage of salt in terms of weight. Since we are dealing with a solution with only two components, water and salt, the difference between the weight of salt and 100 will be the weight of water.

Table 3-1. Depression of the Freezing Point Resulting from Adding 2 g of Chemical (Solute) to 100 g Water (Solvent)

Chemical	$\Delta T(°C)$
Sucrose	0.11
Ethylene glycol	0.60
Urea	0.61
Ethanol (ethyl alcohol)	0.81
Calcium chloride	0.88
Magnesium chloride	1.05
Methanol (methyl alcohol)	1.14
Sodium chloride	1.19

For example, find the number 10 on the bottom scale. This means that 10 percent of the solution we are considering is salt, with the remaining 90 percent of the solution water. This would be the case if we have a tank holding 100 lb (45.36 kg) of salt solution made up of 10 lb (4.54 kg) salt and 90 lb (40.82 kg) water. The vertical scale is a temperature scale. The freezing point of a volume of water that has no salt in it (and thus is 100 percent water), represented by the zero at the bottom left of the chart, is 0°C (32°F). This we can see by following the curve marked "solubility curve" (sometimes called the "liquidus" curve) as it rises to the left until it intersects the vertical temperature scale. The solubility curve is so called because it divides the region where salt is completely in solution from the region where some of the salt will remain undissolved and mix with ice that will form.

There's more that we can learn from this diagram. Adding salt to the same quantity of solution means moving to the right along the horizontal axis. Let's say we increase the quantity of salt to 20 percent by weight (you will frequently see this written as 20 wt%) without increasing the amount of water, so that the proportion of water drops to 80 wt%. As we do so, down goes the freezing point, to about −16.5°C (2°F). We learn this by drawing a line vertically from the 20 percent tick on the bottom (horizontal) scale until it intersects the solubility curve. Now draw a line horizontally to the temperature (vertical scale) and read the temperature on the left or right (vertical) scale. As more salt is added, the freezing point will continue to drop until the salt concentration reaches about 23 percent. Here the freezing point reaches a minimum of about −21°C (−6°F), called the eutectic point. Adding more salt will raise the freezing point. If the temperature is lowered below the eutectic point, salt will no longer remain in solution but will solidify outside the ice boundary and result in a mixture of ice and salt (as $NaCl \cdot 2H_2O$). If, however, we add salt beyond the so-called eutectic composition of 23 percent salt, the freezing temperature rises rapidly, as can be seen by following the solubility line as it rises from the dashed line representing the eutectic temperature. This region of the phase diagram represents a saturated solution, i.e., one that holds the maximum possible amount of solute (salt) at the particular temperature.

The phase diagram for calcium chloride (Fig. 3-4) is similar, but at low concentration the curve is flatter, i.e., the freezing-point depression is less (see Table 3-1). The drop in freezing point increases rapidly as the concentration increases, however, and the curve becomes progressively steeper until the eutectic point is reached at about 30 percent $CaCl_2$ and a freezing point of −67°F (−55°C).

When water freezes, it gives up 334.7 kJ/kg (144 Btu/lb, 80 calories/gram, or 1200 Btu/gal) of heat. This is called the *latent heat of fusion*. Melting is the inverse process, and heat must be added to ice to

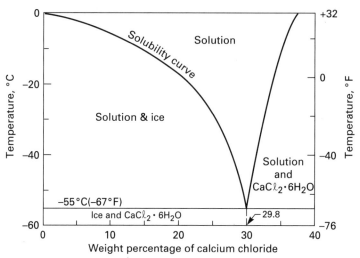

Figure 3-4. Phase diagram for calcium chloride.

melt it. In the case of ice on a pavement, heat can come from the air, from solar radiation, or from the pavement and its heat reservoir. In all cases, however, heat will flow only "downhill," i.e., from a hot to a cold body. When an ice control chemical is spread on ice and goes into solution, the freezing point is lowered, the vapor pressure is lowered, and additional ice tries to reach the more comfortable state of melting. It will do so only if there is a source of heat available.

In summary, these requirements must be satisfied in order to melt ice by using chemicals:

1. Sufficient moisture must be available to dissolve the chemical.
2. A heat source must be available to provide the energy for melting.
3. If complete utilization of the chemical is to be achieved with no waste, the concentration of chemical must not exceed the eutectic composition.

3.2.3 Chemical Selection

Certain chemicals melt ice by the mechanism described in Sec. 3.2.2. This reaction requires only that the vapor pressure of the chemical solution be lower than that of water, and either solid chemicals or aqueous solutions of the chemicals can be used. Some chemicals that have been used for ice control are normally liquid at freezing temperatures and above. The following is a list of chemicals that have been or are being used routinely or experimentally for ice control, grouped into three classes.

1. Solid chemicals
   Calcium chloride
   Calcium magnesium acetate
   Magnesium chloride
   Potassium acetate
   Potassium chloride
   Potassium formate
   Sodium acetate
   Sodium chloride
   Sodium formate
   Urea
2. Aqueous solutions (i.e., dissolved in water) of the normally solid chemicals listed above
3. Liquid chemicals (compounds only found in liquid form)
   Ethylene glycol
   Propylene glycol
   Isopropyl alcohol
   Methyl alcohol

Detailed information on the properties of these chemicals is given in Appendix B. Ethylene glycol is a very effective freezing-point depressant, as has been demonstrated to motorists by its use in automobile antifreezes. It was once commonly used on airport runways both because of its melting effectiveness and because its corrosion of aircraft metals fell within acceptable limits. It has also been applied on highway bridge decks because of low corrosion of reinforcing steel. It is no longer used because of its detrimental environmental effect, resulting from a high BOD (biochemical oxygen demand). Methyl alcohol is very effective in lowering the freezing point of water, but it is toxic in large amounts and has a very high evaporation rate (it also causes blindness if ingested). Isopropyl alcohol is also an effective freezing-point depressant and has been used on airport runways because it too does not corrode metals. High expense has limited its use. Propylene glycol does not have the environmental and toxic drawbacks of ethylene glycol, but it too is very expensive. Several other chemicals in the list above are still in the evaluation stage, with various levels of official approval pending definitive results, but all are included in Appendix B to allow comparison of their properties.

The selection of the most appropriate chemical for a particular use is based on (1) availability, (2) cost (both material cost, cost of special handling equipment, and potential costs for environmental rehabilitation), (3) climatic regime, and (4) performance requirements.

1. *Availability.* There must be sufficient quantities of the chemical commercially available to fill the agency's needs, and there must be an assured resupply capability during the winter.

2. *Cost.* The chemical with the lowest delivered cost may not have the lowest total cost. Corrosion will occur to some extent with all ice control chemicals, as will eventual damage to portland cement concrete. Straight salt has the lowest first cost, but the corrosion it causes can add to its total cost. Where avoidance of corrosion is required, such as on bridge decks, parking garages, or other critical structures, salt with corrosion inhibitors may provide adequate low corrosion rates, but if not, a nonchloride alternative may be necessary. Where there are restrictions on use of salt in the vicinity of public water reservoirs, an alternative chemical will be necessary unless a mechanical control method is practicable and successful. In either case, material cost will be higher.

3. *Climatic regime.* All ice control chemicals are effective in preventing ice formation or in melting any ice that has formed when the temperature is within a degree or two of 32°F (0°C). Even though salt has a eutectic temperature of −6°F (−21°C), its solubility rate is too slow at temperatures below about 23°F (−5°C) to provide fast enough melting for critical roads. When an agency's region has temperatures well below that temperature during much of the wintertime, an alternative chemical will be required for critical highways, or an additive must be used with the salt to lower the freezing point. Addition of calcium chloride to salt, either by mixing solid or liquid calcium chloride with salt during stockpiling or during truck loading or by spraying liquid $CaCl_2$ at the time of spreading, is very effective for two reasons: It initiates the solution of dry salt, and the mixture lowers the effective melting temperature. Magnesium chloride is frequently used as an alternative when a chloride is allowable and the introduction of sodium into the environment must be limited.

4. *Performance requirements.* The main element in this category is the nature of the road system. Critical, heavily traveled routes must be kept operational with a friction level consistent with safety. Steep grades call for special attention if the intent is to keep them fully operational during the winter. Railroad grade crossings, toll plazas, snow routes or evacuation routes, major intersections, and sharp curves also require special attention. Treatment for these locations must be rapid and prompt. Promptness in treating potential problem areas is a mat-

ter for agency scheduling, but rapid chemical action depends on the chemical used, as was discussed in the paragraph above.

3.2.4 Application Methods

The methods and equipment used for applying ice control materials will depend to a large extent on the strategy adopted by the agency. The two principal strategies are anti-icing and deicing. These techniques are covered in greater detail in Sec. 4.11. Included here is a brief summary of the most used methods. Equipment used for material application is described in Chap. 5.

Solid chemicals are distributed on the pavement by either (1) spreading over a wide path (covering parts of two lanes) by means of a spinning disk or a roller extending the width of the truck tailgate, or (2) windrowing in a narrow path [1 to 3 ft (0.3 to 0.9 m) wide] through a tube or off a "dead spinner" (a spinner disk that is not spinning). Concentrated spreading of salt, either in a narrow band on two-lane roads or in a 4-ft- (1.2-m-) wide band spread near the centerline crown or on the high side of superelevations, is favored by many highway maintenance engineers as the most efficient use of chemicals. There are two reasons for this technique: (1) the loss that may result from chemicals bouncing off the road when full-width spread is used is reduced, and (2) a more rapid melting action is achieved by applying a concentrated chemical strip on a snowpack. This will speed the exposure of enough pavement to provide improved traction for at least one set of front and rear wheels, and the concentrated brine will flow under the pack to break the bond, thus enabling traffic and plowing to remove the accumulation. Early exposure of a portion of the road surface to the sun, which may be achieved by concentrated spreading, will also increase the melting rate by absorption of thermal energy. The recent development of so-called zero-velocity spreading equipment provides an alternative method for applying a narrow swath of solid chemical with reduced loss. This is achieved by slinging the material directly behind the truck at a speed equal to that of the vehicle itself. As a result, the chemical strikes the pavement as though it had dropped from a stationary vehicle. Solid chemical loss can also be reduced by prewetting, with other benefits as well. This is described in the next section.

Liquid chemicals are dispensed by conventional nozzle distributor bars attached to the rear of tank trucks or dump trucks carrying a slip-in tank or saddle tanks (tanks, usually formed from plastics such as polyethylene, fitted on the side of the dump body). Nozzles consistent with the range of discharge rates expected to be used should be selected. Fan rather than cone spray nozzles are preferred for best coverage. Some agencies prefer nozzles which provide a narrow stream that will more readily penetrate any accumulation present on the road.

3.2.5 Prewetting

Dry salt spread on a cold, dry snow or ice surface will not immediately begin the melting process. In addition, since the salt particles will not stick to the surface in the dry state, they will readily bounce off the road or will be blown off by traffic action. It takes 3 to 5 min for dry salt to embed in ice at 30°F (-1°C) and 19 min at 25°F (-4°C) (Lemon 1975). Both of these disadvantages to the application of dry salt can be overcome by coating the surface of the dry salt crystals with a film of liquid. Water could be used, but in spite of the attractiveness of this from the material cost point of view, it is much more effective to use a chemical solution. A freezing-point depressant not only adds the liquid "starter" to initiate the melting action by liquefying the salt, but it assists in lowering the temperature of effective melting. Furthermore, water may freeze in exposed tanks on the truck, and adding insulation or heating the water means extra cost. Calcium chloride in liquid form, generally with a concentration in the vicinity of 32 percent, is most effective in lowering the effective melting temperature. However, magnesium chloride and sodium chloride are also quite effective as prewetting agents.

Ideally, each salt grain should be coated with a liquid film when it is applied to the pavement. Three prewetting methods are available: injecting a concentrated solution of calcium chloride (42 percent is commercially available for this purpose) into stockpiled salt, saturation of the salt prior to spreading by spraying a measured amount of solution onto the load in the truck while in the maintenance depot, and spraying the solution onto the salt as it is spread on the road. Truckload prewetting can be accomplished with little specialized equipment—conventional spreading equipment is used. There are several drawbacks, however, including the possibility of incomplete wetting of all the salt and excessive application of the prewetting solution, which may be not only wasteful but environmentally detrimental. Prewetting at the time of application requires application equipment on the truck, consisting of storage tanks, hoses to carry the solution to the point of application, generally a spinner, and a control system to meter precise amounts of the solution to correspond to the distribution rate. For greatest economy and reduction of environmental problems arising from excess chemical application, the salt and the prewetting solution should be spread at a constant areal density regardless of the

Snow and Ice Control Methods

speed of the vehicle. This is accomplished by so-called ground-oriented spreaders. Further description of this equipment can be found in Chap. 5.

Experience in Sweden has demonstrated that it is most effective to wet salt at the rate of 14 gal/100 lb (80 to 100 L H_2O/t) salt and apply at the rate of 40 to 234 lb/la-mi (3 to 18 g/m^2). This mix can be effective down to $-12°C$ (10°F), although $-6°C$ (21°F) is usually used as the lower decision value. For comparison, the application rate of liquid salt is 20 g/m^2 (260 lb/la-mi), equivalent to 5 g/m^2 (65 lb/la-mi) of dry salt.

3.2.6 Application Rates

Determination of the proper application rate is a matter of judgment and an educated guess as to what weather conditions immediately following the application will be. This does not mean that the choice is completely wild, since a number of very effective tools have been developed as a result of research and intelligent experimentation by highway agencies. Improved application equipment is capable of delivering a precise quantity of chemical chosen to achieve a desired effect. The optimum application rate depends on the

- Level of service required
- Weather conditions and their change with time
- Form (whether liquid or solid) and characteristics of the chemicals used
- Time of application
- Traffic density at the time of, and subsequent to, chemical application
- Topography and the type of road surface

More detailed guidance for application rates is given in Chap. 4.

3.2.7 Storage and Handling of Materials

Materials used for ice control are meant to be applied to roads in controlled amounts to accomplish a specific objective. It's not overstating the obvious to point out that any substance can be harmful in excess—even water (it's called drowning). For this reason, and for reasons of sound cost control, chemicals in particular should be stored to prevent unnecessary loss into their surroundings, where they can enter the environment and ultimately the food chain. The most commonly used ice control chemical, salt (sodium chloride), is very soluble in water, and therefore much of this discussion will revolve around its storage. Covered storage is the minimum control method that should be used for any chemical or for any material containing soluble chemicals, such as abrasive piles. Tests in England have shown that the loss from a large (1000-ton) uncovered salt stockpile exposed to rain will amount to 0.0013 ton/ton•in (0.000512 kg/kg•cm) of rainfall (Hogbin 1966). As an example, a 1000-ton pile on which 15 in of rain falls will lose 19.5 tons, or nearly 2 percent. The leachate is almost a saturated solution, thus the loss is proportional to the rainfall times the base area of the pile. In addition to preventing loss, keeping precipitation from reaching stockpiles will reduce the formation of lumps or a wet, caked mass which can interfere with loading and spreading.

Storage Methods. It has been the practice to store large piles of salt on impermeable pads and to cover them with waterproof sheeting secured in place with weights such as old tires lashed together with ropes that are tied to some anchor. This is a satisfactory method when the cover is rolled back for loading only in dry weather. It is not a desirable method for use during periods of precipitation when spreader trucks must be loaded from the pile. For that purpose, some type of building must be provided (see Fig. 3-5a–d). The storage capacity needed to main-

Figure 3-5. Storage facilities for salt: (a) Simple roofed structure.

Snow and Ice Control Methods

Figure 3-5. (*continued*) Storage facilities for salt: (b) dome structure; (c) silo.

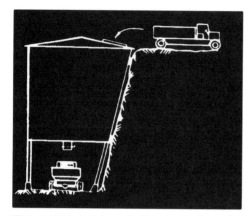

Figure 3-5. (*continued*) Storage facilities for salt: (*d*) hopper adjoining a hill.

tain a ready supply can be based on the historical consumption rate during the winter and frequency of reordering from the supplier during the season. Many designs of buildings are used for storage of these active stockpiles, from simple sheds with fixed roofs, to sheds with slide-back roofs, to large dome-shaped structures. Salt has an angle of repose of 32° (see Fig. 3-6). This means that an unconfined pile will form a cone with sides that slope at an angle from the horizontal of 32°. Floor area for such a pile can be based on a bulk density of 80 lb/ft^3 (1280 kg/m^3). Piles can be built higher by using containment walls. These must be capable of withstanding the horizontal load of the pile. In all cases the floor should be paved. Ground-level storage facilities will require some device for loading mechanized spreader trucks, such as a front-end loader. Storage in silos will enable loading by gravity, as will hoppers set against a hill so

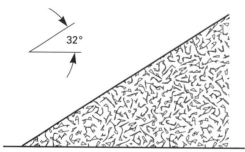

Figure 3-6. Storage area required for an unconfined salt pile is based on an angle of repose of 32°.

that the hopper can also be filled by gravity. Other best practices include selecting a site located away from wells, water supply reservoirs, and groundwater sources; paving the loading area; and installing a system for collecting any chemical runoff from the storage and loading area.

Containment of Runoff. There are only three ways to get rid of brine that has collected: (1) evaporation, (2) pumping it to some disposal location, or (3) applying it on roads. Evaporation will require a pond sufficient to hold the anticipated quantity of runoff for a duration dependent on the evaporation rate for your area (Pletan 1972). Pumping the brine to a disposal area may involve loading it into tank trucks for transport to an approved disposal area, which may be a river during periods of peak flow when dilution will be sufficient. If most of the leachate is collected during the winter, it may be used directly for application on roads. However, if most runoff occurs during the summer, it may be impractical to hold a large amount until the following winter, so disposal will be necessary.

Salt vendors can provide more detailed instructions for designs of buildings and recommendations for good storage practices.

3.2.8 Environmental/ Infrastructure Effects

Corrosion. Corrosion is the destructive attack of a metal in the presence of moisture by chemical or electrochemical processes (Hospadaruk 1978). Metals used as construction materials are refined from basic ores using large amounts of energy, and when they are exposed to a natural environment, they will eventually revert to their original, natural state. This is a natural process which can only be slowed down. Corrosion occurs when an electrochemical cell is established, with oxidation of the metal at one site (anodic) and reduction occurring at a nearby (cathodic) site. A flow of electrons ensues, called the *corrosion current*. The corrosion rate is directly proportional to the magnitude of the corrosion current. Disrupting the current flow, therefore, is the key to reducing corrosion.

Many ice control chemicals contribute to corrosion of metals in three principal ways: (1) by increasing the conductivity of solutions, thereby increasing the corrosion current, (2) by prolonging the time that surfaces remain wet and therefore contributing to corrosion current flow, and (3) by providing ions such as chloride that aggressively penetrate and destroy protective oxide films on metal surfaces (Chance 1974). Ice control chemicals are not the only culprits contributing to corrosion of vehicles and structural elements, however. Salt carried inland from marine environments is another source of corrosion, though certainly not to as great an extent as is road salt.

In light of the mechanisms enumerated in the previous paragraph, corrosion can be reduced by using a chemical that provides fewer charged particles (ions) for the corrosion current, by preventing intrusion of water or its retention in the vicinity of a metal surface, or by introducing a barrier to the migration of water to the metal surface. The highly ionized chemicals such as sodium chloride, calcium chloride, and magnesium chloride dissociate completely and thereby provide many charged particles for the corrosion current. Organic chemicals such as calcium magnesium acetate, sodium formate, and urea dissociate into fewer charge-carrying ions for an equivalent mass of material, so the corrosion current may be lower. Sealing the surface of portland cement concrete (pcc) so that water is less likely to penetrate and carry the corrosion-accelerating chemicals to the steel reinforcing bars is another method to reduce corrosion. Application of linseed oil to well-cured pcc is a technique that has been used successfully. Coating steel with an impermeable film such as epoxy resin is the third method for reducing the potential for corrosion.

Scaling of pcc pavements is a likely consequence of the application of ice control chemicals. Though not properly termed corrosion, it does represent physical damage. It is the result of a physical action rather than of chemical attack of the cement matrix. Cycles of wet-dry and freeze-thaw generate expansionary forces in the concrete which can lead to spalling. Various hypotheses have been advanced regarding the mechanism of the process of deterioration, such as the presence of a chemical concentration gradient, continuous contact with the chemical solution, osmotic pressure, thermal shock, and drying shrinkage (Browne et al. 1970). Regardless of the mechanism, experience has demonstrated the effectiveness of several techniques to reduce deterioration: (1) Cure pcc for at least a year before application of ice control chemicals, (2) use an air-entraining agent, and (3) seal the surface. Many proprietary materials for sealing portland cement concrete are available, and one of the simplest, cheapest, and perhaps oldest is boiled linseed oil mixed with mineral spirits. Performance tests of several surface coating treatments can be found in Pfeifer and Scali (1981).

Vegetation. Any substance used in quantities in excess of the natural environmental load will either temporarily or permanently affect that environment and its inhabitants. This applies to water as well as to the chemicals applied for snow and ice control. The effects of chemicals on the environment fall into two classes: stimulation of natural processes and degradation of the natural environment. Chemicals whose effects are in the first category are those that speed up growth processes. This includes urea, a nitrogen-rich fertilizer which has been widely used for ice control because of its low potential for corrosion. On the downside, it speeds up vegetative growth and in large quantities will kill vegetation. This group also includes phosphate-containing compounds such as tetrapotassium pyrophosphate. These nitrogen and phosphate compounds stimulate the growth of algae in lakes and streams, which depletes the oxygen dissolved in the water and leads to loss of aquatic life, a process called eutrophication. Most ice control chemicals fall into the second category. The most conspicuous member of this group is salt. It's a bad actor because of corrosion, of course, but the sodium ion, which makes up about 39 percent (by weight), can also displace the potassium, calcium, and magnesium in clays and thereby adversely affect the soil structure. Salt can also injure roadside vegetation by increasing the soil salt concentration, which can result in salt absorption through roots that creates osmotic stress. Direct contact on leaves, needles, and branches resulting from splash and spray can also lead to decline of plant life.

Salt-Tolerant Plants. One approach to reducing the damage to vegetation that can result from the use of salt is to replace sensitive plants with salt-tolerant plants. In many areas of the world, highly saline soils support vigorous vegetative growth. Agronomists have cataloged the species of bushes, shrubs, and trees that will tolerate high levels of salt and still remain healthy. A list can be found in Appendix B.

3.2.9 Chemically Impregnated Pavement

Dispersing a freezing-point depressant within the wear course of a pavement can provide a continuous supply of bond-preventing chemical to the surface. That is the principle behind a commercial product called Verglimit. Calcium chloride pellets encapsulated in linseed oil are added to asphalt concrete during mixing in the batch plant. After this concrete is placed on the existing pavement as an overlay, sand is spread on the surface to provide a temporary increase in friction. Normal wear of the surface exposes particles of the calcium chloride, which absorb moisture from the air and dissolve. The result is an always-present thin film of chemical which in winter will inhibit ice-pavement bond formation. Of course, the thin film of chemical will also be present during snow-free seasons and consequently can reduce friction to a small extent. Careful installation is necessary to avoid creating extremely slippery surfaces. Laboratory tests have indicated that addition of calcium chloride to the asphalt concrete improves temperature susceptibility mainly by increasing the resistance to rutting at high temperatures. Verglimit increases the susceptibility to moisture damage, as measured by retained tensile strength and resilient modulus ratios, because the particles absorb water and the specimens swell. However, there is a decrease in the amount of stripping as determined visually (Stuart and Mogawer 1991).

3.2.10 Problems with Chemicals and How to Avoid Them

Chemical freezing-point depressants have become an essential tool in winter maintenance, and judicious selection and use will reduce the problems described above to a level acceptable to most reasonable people who are also motorists. If corrosion is the major problem, a nonchloride chemical is the obvious choice. If protection of vegetation is the objective, avoiding high-nitrogen compounds and sodium-containing chemicals is the approach to take. If reduction of sodium intrusion into water bodies used for domestic water supplies is required, low-sodium-content chemicals will be the choice. In all cases, application of the minimum amount of chemical is an obvious requirement. Gone are the days when some municipalities applied as much as 1000 lb/la-mi (2818 kg/la-km) to "burn off" the snow so that plowing would not be necessary. Recent experience has demonstrated the effectiveness of anti-icing in reducing the quantities of chemicals applied to roads to maintain them at, or to return them to, an acceptable friction level. This is in contrast to the more conventional practice of deicing roads, where snow or ice are allowed to accumulate on a pavement to a specified depth [often 2 in (5 cm) of snow] before any treatment is commenced. The deicing of compacted snow that generally becomes necessary when this practice is used requires greater quantities of chemicals to return the pavement to an acceptable friction state. More detailed discussion of anti-icing can be found in Chap. 4.

3.3 Mechanical Methods

Into this category fall plowing, high-velocity or high-volume air blowers, and brushes or brooms. The objective of mechanical removal is to pick up the snow from the road, shearing it from the road if necessary, and cast it to a storage area off the road. Plowing is most commonly accomplished by displacement plows mounted on the front, side, or beneath their truck carriers, or by rotary plows which pull the snow into a rotating element and cast it to the side. Mobile air blowers are used on airport runways, either alone or in conjunction with powered brooms. Air with a velocity high enough to dislodge lightly bonded snow and blow it several hundred feet is directed at an angle nearly parallel to the pavement and very close to it. It is not unusual for the high-velocity air stream to pick up pieces of pavement if there are wide cracks. Powered brooms must move relatively slowly over the pavement for efficient removal. The low speed is acceptable on airport runways, but not for use on most roads. However, work is under way to design brooms that can be operated at speeds acceptable for highway operations.

Ice is weak in bending and will readily fracture. This is the principle of a little-used mechanical method: incorporating rubber granules in the wear course of asphalt concrete to provide a surface that deforms under load. Ice that is brittle at low temperatures will crack when its support is flexed under wheel loads and may disbond from the pavement surface. A material called PlusRide, which originated in Scandinavia as Rubit, is commercially available to provide the deformable properties. Laboratory tests have shown that PlusRide reduces the stiffness of the mixtures and increases the amount of permanent deformation at all temperatures, thereby increasing the resistance to low-temperature cracking but decreasing resistance to rutting. The material has a variable effect on moisture susceptibility; in some cases it may increase the retained tensile strength and resilient modulus ratios and decrease the amount of swelling which occurs when conditioning the specimens in water. In other cases, it may decrease the retained tensile strength and resilient modulus ratios and increase the amount of swelling during conditioning (Stuart and Mogawer 1991). The experience of several agencies that have installed sections of deformable pavement as an ice control measure has been that the method was not effective enough to justify the added cost.

3.3.1 Distinction between Snow Removal and Ice Removal

Frequently no distinction is drawn between the two winter maintenance activities. There are, however, several differences between the two activities which can influence strategy, materials use, and equipment. Chief among them is the difference in the volume of material that must be removed. Even light snowfalls require a tall plow to prevent snow from blowing over the top and being redeposited on the road. Blow-over will also reduce the driver's visibility. Ice, because of its much smaller amount and its higher density, can be plowed with a low blade such as on a grader or the underbody blade on a truck. Another distinction is the effort required to shear the precipitation from the pavement. New snow that has not been trafficked by vehicles will not be tightly bonded to the pavement and can readily be picked up by the plow cutting edge. Ice, on the other hand, will bond to the pavement with a range of strengths; at temperatures close to the melting point of water, the bond may be broken by the plow cutting edge with ease, but as the temperature drops, the bond increases in strength. The ice-pavement interfacial bond strength may even exceed the tensile strength of the pavement, in which case failure during removal may occur within the pavement itself. Removal by mechanical means becomes very difficult and often impossible, even with many passes of the blade. Weakening the ice-pavement bond by heat or by a chemical freezing-point depressant becomes necessary. Snow that has been compacted by traffic may bond to the pavement as strongly as ice, and though a blade scraping its surface may cut off some material, the bond may not be broken and much snow will remain. Removal must employ methods similar to those used for ice.

3.3.2 Equipment for Snow and Ice Removal

See Chap. 5, "Snow and Ice Control Equipment," and Chap. 9, "Trucks For Snow and Ice Control."

3.4 Thermal Methods

The purpose of heating a pavement is to reduce traffic delays, personal injury, and property damage from accidents caused by black ice, glaze ice, and/or packed snow and to do so expeditiously at an acceptable cost. The cost of installing a fixed system and operating it, or of making and operating a mobile heating apparatus, is too high for general use. Bridge decks are prime candidates for heating, since their undersurfaces are exposed to cold air, allowing ice to form earlier than on the pavement on grade. Pavement there receives heat from the ground until exhausted as winter progresses. Preferential bridge deck icing, as it is called, is a major problem in many areas of the country where the combination of calm air, nighttime radiation cooling, and a source of water vapor below the bridge produces black ice. This is usually a very transient condition, frequently lasting for only a matter of minutes, but it is extremely hazardous until it melts. Though preemptive chemical treatments are gaining favor for combating many of these and similar icing situations, heating a pavement is a positive control method. In addition to bridges, critical locations where heating systems have been installed include toll plazas, on and off ramps, and steep grades. Airports have immense paved areas which would be costly to heat in their entirety, but research is under way to provide heated surfaces for critical locations such as touchdown areas, ramps, and some taxiways.

Snow and Ice Control Methods

One of the unusual properties of water is that it can be supercooled, i.e., it can be made to remain in the liquid state even at several degrees below its normal freezing temperature of 32°F (0°C) under special conditions. Also unusual is another interesting fact: Though ice can exist in equilibrium with water at 32°F (0°C), it cannot exist in a superheated state, i.e., at temperatures above 32°F (0°C). What is the significance of this? As long as a surface is maintained at a temperature of 32°F (0°C) or above, ice will not form or, if initially present, will melt.

3.4.1 What Are Thermal Methods?

How can the temperature of a pavement be raised above the freezing point of water if it is initially below that temperature? Or how can a pavement be maintained at a temperature above 32°F (0°C)? There are only two basic methods: Add thermal energy from above the pavement, or add it within the pavement itself. The basic laws of physics dictate that heat can be transferred by three mechanisms, singly or in combination: conduction, convection, and radiation. Conduction requires contact between the heat source and the object being heated. That is the only way that heat can be added within the pavement, by incorporating heat sources in the pavement itself. Adding from above, however, can be achieved using any of the three mechanisms, though conduction is impractical because of the length of time a heat source must be left in contact with the cold surface before sufficient energy is transferred to raise the temperature even slightly. That leaves convection and radiation as the mechanisms for adding heat from above the surface. Convective heat transfer alone is a rather slow process, so again the combination of radiation and convection is necessary for a possible approach. Use of radiation alone is also a possible method; this is discussed later in this section.

Heated Pavements. It requires a large quantity of heat to melt snow and ice. For example, to melt 3 in (8 cm) of new snow on just one lane-mile of road will require the energy equivalent of nearly 95 gal of fuel oil, assuming no losses (the equivalent for 1 la-km is 223 L). Multiply that by the hundreds or thousands of lane-miles in a city, county, or state, and it's clearly uneconomic to attempt complete removal of large snow accumulations using heat alone. The value of heating lies in the elimination of light "nuisance" snowfalls and thin ice accumulations that otherwise might require the dispatching of a spreader truck to apply chemical or abrasive. In addition, pavement heating can prevent snow or ice from bonding to the pavement, enabling more complete removal by plowing. In practice, therefore, heated pavements are justified only where a critical situation exists, such as bridge decks, steep grades, railroad crossings, and toll plazas.

One of the most effective methods of raising the temperature of a pavement to a level at which ice cannot form or remain is to add thermal energy into the pavement itself (see Fig. 3-7). The means for conveying the heat into the pavement can be a fluid, either air or a liquid, or it can be electricity. Active processes require a source of energy to pump the heat into the pavement. A device called a heat pipe is a passive method, i.e., it requires no external energy source for extracting the heat from a source and delivering it to the pavement, as will be described below. Since air has a low heat capacity, only about 1/400 that of water, it is not a useful fluid for heating a pavement. (However, there is an interesting use of air for conditioning a large mass: Some buildings in the Arctic, such as hangars, must be built on slabs directly on the ground because of the heavy loads on them, and in order to keep building heat from melting the permafrost below the foundation, ducts are opened in the winter to cool the subgrade well below freezing, then closed in the short summer to prevent warmer air from entering.) Water has a very high heat capacity and is used with antifreeze additives as a heat transfer medium in so-called hydronic systems.

Hydronic Systems. In these systems, a fluid is heated and circulated through a network of tubes or pipes buried close to the surface of the pavement. The pipes are laid in a transverse rather than longitudi-

Figure 3-7. Heated pavement with embedded pipes carrying hot fluid.

nal pattern. Materials used for pipes are metals (wrought iron, carbon steel, copper, or stainless steel), and those used for tubes are synthetic elastomers such as nylon or EPDM (ethylene propylene diene monomer). If plastic is used rather than carbon steel, the temperature drop between the inside and outside walls will increase by a factor of 200 for the same size pipe, thereby requiring a higher fluid temperature. Water has a very high heat capacity, and so water with some type of antifreeze added is generally used as the working fluid for these systems. Propylene glycol is commonly added as the antifreeze. Some installations use an oil-based heat transfer fluid. Heat is generated by conventional sources such as furnaces burning fuel oil, propane, or natural gas. Corrosion of steel pipe is a potential problem if chloride ice-control chemicals will be applied or tracked on the heated pavement. This can be reduced by ensuring that at least a 2-in (5-cm) cover of portland cement concrete is placed over the pipes. Thermal stresses induced in the pipes can disbond the pipe from the concrete cover unless concrete completely surrounds the pipe; $1\frac{1}{2}$ in (3.8 cm) of concrete should be allowed underneath the pipe. If thermal stresses in the pipe are not resisted by the pipe-concrete bond, pipe failure may occur. Coating the steel pipe with epoxy will also add protection.

Heat Pipe. A heat pipe* is a device which transfers heat by boiling a fluid at one point and condensing it to liberate its heat at another. It consists of a sealed tube or chamber, which may have different shapes, whose inner surfaces are lined with a porous capillary wick. The wick is saturated with the liquid phase of a working fluid, and the remaining volume of the tube contains the vapor phase. Heat applied at the evaporator end by an external source vaporizes the working fluid in that section. The resulting difference in pressure drives vapor from the evaporator to the condenser, where it condenses, releasing the latent heat of vaporization to the heat sink in that section of the pipe. Depletion of liquid by evaporation causes the liquid-vapor interface in the evaporator to enter into the wick surface, and a capillary pressure is developed there. This capillary pressure pumps the condensed liquid back to the evaporator for reevaporation. The process will continue as long as the flow passage for the working fluid is not blocked and a sufficient capillary pressure is maintained (Chi 1976).

Heat pipes are most commonly used to tap geothermal heat. In such an installation, the evaporator ends of the heat pipes are buried in the

---

*The term *heat pipe* was first used in 1963 by G. M. Grover of Los Alamos, who obtained U.S. Patent 3,229,759 for "Evaporation-condensation heat transfer device." Heat pipes were first suggested by R. S. Gaugler in 1942. Early predecessors, "Perkins pipes," existed in the late nineteenth century (Ivanovskii et al. 1982). A Perkins pipe is similar to a heat pipe except that it does not have a wick, which limits use to situations where gravity will return the condensate to the evaporator section.

ground at a depth of about 30 ft (9 m) (where the temperature approaches the mean annual temperature of the area). The condenser ends are embedded in the pavement. That describes the operation in winter, when heat is extracted from the ground and released in the pavement. Bending the pipe from vertical in the ground to horizontal in the pavement has little effect on the performance. There are several installations in the United States making use of geothermal-sourced heat pipes for pavement heating, one of which has been in operation, completely untended, for 20 years and has never required maintenance. The attractiveness of this device lies in its passive operation: Once it is installed, no energy need be supplied. Where a geothermal source is not available, heat to drive the vapor transfer in a heat pipe can be provided by conventional fossil fuel boilers.

Electrically Heated Pavements. Two methods are available for heating a pavement using electric energy: resistance cables and conductive pavement. The resistance to flow of electricity in any conductor generates heat, a physical fact we experience every day (in the heating of an electric light bulb, for example). The greater the resistance to flow of electricity, the greater the generation of heat, but also the greater the "push," the voltage, that is needed to enable the flow. The engineering of an installation requires consideration of the maximum allowable voltage, the resistance of the material, the spacing of the electrical cables, and the heat required at the surface of the pavement. The information presented here is only a brief overview, and any installation will need competent engineering design.

Mineral-Insulated (MI) Cable. MI cable consists of one or two resistance heating elements encased in highly compressed magnesium oxide and wrapped in a copper or stainless steel sheath. The application requirements will determine the necessary cable resistance and sheath material. A high-density polyethylene (HDPE) sheath is added to obtain the greatest mechanical protection and protection from corrosion as a result of penetration of water and chemicals. This also protects the cable during installation. A single-element cable loop must be returned to the power source to complete the circuit. A two-element cable does not have to be looped to the source, but instead can be installed as a stub or pigtail, since the two conductors provide the return circuit path. A cable bend cannot have a diameter less than about 3 in (7.6 cm) to avoid cracking the insulation.

Conductive Pavement. Resistance heating cables or hydronic tubes or pipes are laid out in a grid, with spacing between these linear heat sources varying from a few inches to a few feet. Spacing is based on the heat output of the cable, tube, or pipe; the thermal conductivity of the surrounding paving material; and the design temperature of the surface. The design temperature is based on the design objective: either ice control (prevent the formation of thin films of ice) or snow melting (melt a specified quantity of snow in a specified time over some range of temperatures). Pavement in the immediate vicinity of the heat sources will warm more quickly than the areas between them, creating a thermal gradient. Since conventional pavement materials have low thermal conductivity, reducing the lag in raising the entire surface to the design temperature requires heating the heat sources to above the design temperature. However, if the entire pavement surface is a heating element, this lag and overtemperature can be avoided. That is the purpose of a conductive pavement. Normally, pavement materials, whether asphalt or portland cement concrete, do not conduct electricity and therefore cannot be active resistance heaters. However, if a calculated quantity of conductive particles is mixed into the pavement materials during their construction, electricity can be passed through the material with a resistance high enough to produce an effective heat level. Both conductive asphalt and conductive concrete are being evaluated for use on bridges and airports.

Temperature Control. The simplest heated pavement installation, for a limited area such as a sidewalk, garage ramp, or truck loading area, could dispense with any automatic control and instead be turned on and off manually. The output for which the system is designed would produce the heat necessary for the expected temperature and precipitation conditions. Large or critical installations, regardless of the method by which the heat is produced and delivered, need automatic controls for remote, responsive, cost-effective operation. Three stages of control are necessary: anticipatory control for initiating the heating cycle before the need arises based on air temperature, temperature trend, dew point, and precipitation; idling control for maintaining the heat at a low level between precipitation periods; and operational control for maximum design heat output during a precipitation event. Temperature sensors must be installed at several locations within the pavement to provide real-time measurements to the automatic controller or to the manual control operator.

Direct Radiation. As mentioned above, the smaller the mass that is heated, the less the thermal energy required. Ideally, if just the ice-pavement interface could be heated to the melting point, the ice would no longer adhere to the pavement and could be removed mechanically before it has a chance to refreeze. Ice, however, is a strong absorber of radiant energy, with the result that when heat is added from above, the entire mass of ice absorbs the energy and melts. Ice and compacted snow have a tremendous capacity for absorbing heat, so melting the entire ice layer or snowpack is economically impractical on a large scale. The sun, of course, can do the job free of charge, but unfortunately it isn't always shining when it is needed. The latent heat of fusion—the amount of energy required to melt ice—is 144 Btu/lb (335 kJ/kg). In practical terms, that means to melt $\frac{1}{8}$ in of 27°F (3 mm of −5°C) ice on one lane-mile (1.6 la-km) of road will require about 95,000 Btu (100 MJ). This is the heat equivalent of nearly $\frac{3}{4}$ gal (2.8 L) of gasoline.

Heat sources such as flames or heating elements raised to a glowing red temperature emit their thermal energy in that portion of the electromagnetic spectrum called the infrared, with wavelengths from around 1 μm to 1 mm (10^{-6} to 10^{-3} m) (1μm = 0.00004 in) (Fig. 3-8). Ice absorbs energy in that region nearly completely, but it is transparent to energy in certain other parts of the spectrum. It is more transparent, for example, in some portions of the microwave region, which has longer wavelengths, from 1 mm to around 100 mm. The use of microwaves to break the ice-pavement bond has been investigated, but no practicable device has yet been developed which can generate and apply sufficient energy rapidly at acceptable cost. Infrared energy is used for melting snow and ice on small areas such as building entrances and loading platforms. Infrared luminaries powered by either electricity or natural gas are available for this purpose.

3.4.2 Design Factors

There are two objectives to heating a pavement: (1) providing sufficient heat at the pavement surface to melt all frozen precipitation as it falls, or at least to allow little if any accumulation during an "average" snowfall so that no mechanical removal is necessary, or (2) operating at a heat output level that will prevent ice/compacted snow bonding so that mechanical removal can clear most of the snow or ice and the residue will be melted. The purpose in either case is to maintain an ice-free surface following the precipitation event. Heat outputs of 170 to 275 Btu/ft²·h (536 to 867 W/m²) are required in the coldest locations to melt snow as it falls at a rate of 1 in/h (2.5 cm/h), and a similar snowfall rate over a 24-h period with snow allowed to accumulate to some extent but melting the entire amount by the end of the storm, requires about 90 Btu/ft²·h (284 W/m²) (Williams 1976). It should be kept in mind that thermal methods may be inappropriate in regions with persistent very low temperatures because cold snow will not bond tightly to cold pavement, and freezing rain is not generally experienced. The heated surface should register the same temperature at a given time over the entire area, unless specific design features such as increased heat output in drainage channels to ensure complete drainage of meltwater have been incorporated. The time required between energizing the heated pavement and reaching the maximum pavement temperature should be determined. This will help establish the minimum time for activating the heating system prior to a

Electromagnetic Spectrum

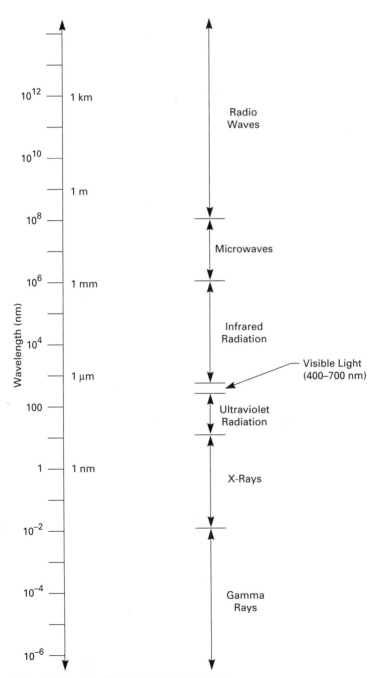

Figure 3-8. The electromagnetic spectrum.

snow/ice control action. A baseline measurement under snow-free conditions is necessary. An installation should be tested before it can be depended upon for operational use. Suggested details to check include the uniformity of heat distribution (determined by infrared scan), the time required to reach 33°F (1°C) from a cold start, and the response of sensors for detecting the presence of ice or snow.

3.4.3 Disposal of Snow; Snow Melters

The large volumes of snow cleared from roads and walkways in built-up areas of cities, industrial and warehouse clusters, extensive parking lots, exposed multistory parking structures, and airport ramps may exceed the storage capacity of untrafficked areas and therefore will require disposal by some means. Loading trucks using front-end bucket loaders or rotary plows is the most common solution, but if disposal sites are distant and trucking costs are high, or if the congestion resulting from the use of many trucks or loaders is unacceptable, it may be cost-effective to use sewer disposal. Where there is sufficient flow in the sewer system and the volume of snow is not too great, dumping the snow into sewers is a practicable solution; in Sapporo (Hokkaido), Japan, a channel has been constructed on the side of an urban street in a densely populated neighborhood, and river water diverted to it carries away snow dumped in it. Generally, however, the snow must be melted or reduced to a slush before it is discharged into a sewer system. Fixed melting pits are used on some airport ramps (Logan International in Boston, Massachusetts, for example) and have been installed in some cities, notably in eastern Canada. Mobile snow melters that can be transported to a problem area have also been used. Both fixed and mobile snow melters commonly use direct-fired oil burners, directing the hot combustion gases through pipes in the melting pit. A design incorporating "submerged combustion," whereby heat transfer is enhanced by discharging the combustion gases into the melting snow, was installed in some of the later systems. Disposal by melting and discharging into the sewer has declined in use because of the steep increase in fuel cost since the systems were originally installed.

3.5 Control of Drifting Snow

Reducing the amount of snow blowing onto a road and forming drifts may have two benefits: enhancing safety and reducing the task of mechanical removal. Obstacles upwind of a road may create "finger" drifts, narrow intermittent drifts that may be unexpected or difficult to see and that can cause a motorist to lose control of the vehicle. Visibility may be reduced in blowing snow conditions and lead to accidents. Snow blowing onto the road may melt and refreeze, causing dangerous patchy ice. Maintenance costs can be reduced significantly if there is less snow to remove. Snowdrift accumulations that create major traffic obstacles or require extra effort to clear frequently recur year after year in some problem locations. These become well known to maintenance crews and are prime candidates for installation of a control barrier. There are three types of control barriers: snow fences, "living snow fences," and expedient vegetative barriers. Of course, no barrier placed upwind of a road will prevent snow accumulation during low wind conditions.

3.5.1 Snow Fences

Fences of one type or another have been used in Scandinavia since the mid-nineteenth century, and in the United States since the construction of the transcontinental railroad shortly thereafter. Large stones and split rail fences were among the types tried, leading to the common use of the 4-ft- (1.4-m-) high picket fence made of narrow vertical slats extending to the ground. Its poor performance, owing to low storage capacity and poor snow collection, discouraged widespread use. Modern snow fence designs are based on field and laboratory aerodynamic studies, and are engineered for peak performance based on detailed knowledge of site characteristics. Some designs take advantage of modern materials such as plastic sheeting which speed up installation and reduce material and labor costs.

Design Factors. Following is a brief review of the important criteria for design of a high-performance snow fence. Fuller detail on design can be found in two publications produced by the Strategic Highway Research Program: *Snow Fence Guide* (Tabler 1991), a 61-page pocket-size introduction to snow fence technology, and *Design Guidelines for the Control of Blowing and Drifting Snow* (Tabler 1994), a 364-page manual intended for design engineering staff.

New snowflakes falling when no wind is blowing will lazily float down to Earth, attracted by its gravitational force. Another force is necessary to lift the snow crystals and carry them once more into the air. At a certain speed, wind will do this. Depending on the sizes of snow crystals, their exposure, and how well stuck together they may be, a wind of around 10 mi/h (4.5 m/s) will begin to pick up some of the particles. As the wind speed increases, greater quantities of snow will be suspended and the snow will be carried higher. The purpose of a barrier is to lower the wind speed, thereby reducing the wind's snow suspension capability. A well-designed barrier will accomplish this efficiently and will remain effective over most, if not all, of a season's snowfall.

To remain effective over an entire season, a snow fence must be designed with sufficient capacity. The total amount of snow that can be expected to blow past the point where a fence system is planned must be estimated. This involves the distance, called the *fetch*, over which the wind can pick up the snow, and the amount of snow available during the season to be picked up, termed *relocated precipitation*. A practicable estimate of this amount is 7 percent of the annual snowfall for the area. The combination of fetch and available precipitation will determine the required storage capacity and therefore the fence height. Factored into the design process are the losses from evaporation, consolidation, and melting. Height is the most important consideration because it determines the storage capacity and has a great influence on trapping efficacy. As an example of the importance of height, adding 6 in (15 cm) to a 4-ft (1.2-m) fence, an increase of $12\frac{1}{2}$ percent, will increase capacity by 30 percent. Trapping efficiency will also be strongly affected by the gap between fence bottom and ground. The optimum bottom gap is 10 to 12 percent of the total vertical height (many fences are constructed leaning into the wind at an angle of about 15° from the vertical, so the total vertical height is slightly less than the fence height—in this case, $\cos 15° = 0.97$ as high). A solid fence will quickly become saturated, i.e., reach its capacity and cease to act as an effective collector. Tests have shown that a solid barrier has only 33 percent of the capacity of a porous fence of equal height. Openings in the fence, the porosity, amounting to 40 to 50 percent of the face area are most effective.

Steps in the design of a fence system include the following:

- Determine the predominant wind direction.
- Estimate seasonal snow passing the fence location.
- Determine fence height and number of rows of fences to contain this amount of seasonal snowfall.
- Determine fence placement to intercept snow before it reaches the area to be protected.
- Select construction design and material.

Fence Placement. A fence should be placed far enough from the protected area so that the downwind drift does not extend into it. On flat terrain, the minimum setback distance for a 50 percent porous fence is 35 times the fence height. Going well beyond this distance will reduce the

Figure 3-9. The Wyoming truss–type fence.

protection because snow can be picked up downwind of the fence and carried onto the protected area. It may not be possible to choose the optimum distance because of terrain features, but modifications of height and use of multiple rows may be acceptable alternatives.

The Wyoming truss–type fence (Fig. 3-9) is an example of a commonly used design consisting of horizontal 1-×6-in. (2.5-×15-cm) wooden boards fastened to wooden trusses, anchored by steel reinforcing bars driven into the ground. Average porosity is 45 percent, bottom gap is 10 to 12 percent of total height, tilt into the wind is 15°, and panel length is 16 ft (4.9 m). Fences of this design can withstand winds up to 100 mi/h (45 m/s) and snow settlement forces arising from complete burial. Horizontal rails are best because a bottom gap is likely to be retained even after snow accumulates at the fence bottom. As long as a bottom gap remains open, however, there is no difference in snow storage capacity between horizontal and vertical openings for materials having 40 to 50 percent porosity.

Fence Materials. Metal, plastic, and woven fabrics, as well as wood, have been used successfully as fencing material. Most plastic fencing is made of polyethylene. Polypropylene, fluorocarbons, and EPDM are also used. Tensile strength, resistance to ultraviolet degradation, and size of openings are the important criteria for selection.

Posts for supporting fencing materials are made of either wood or steel. They must be strong enough to withstand the high forces that will occur at wind speeds of 100 mi/h (45 m/s). Plastic fencing must be kept taut between posts, and this requires tensions as high as 250 lb/ft (3.65 kN/m) of height.

3.5.2 Living Snow Fences

Living snow fences are vegetative barriers that control drifting snow in the same manner as wood or plastic fences and can be equally effective. Trees, shrubs, grass, and standing corn stalks or sunflowers have all been used. Though the same principles apply to living snow fences as to structural fences, guidelines for their location and siting are modified to account for porosity and height changes. As the living barrier becomes more dense (i.e., less porous), more snow is stored in the upwind drift and the downwind drift becomes shorter. Countering this, however, is the increase in height, which tends to make the drift longer. The distance of the living fence from the road must be sufficiently great to prevent snow encroachment on the road as the plants grow. It may take several years for a new living snow fence to become effective. During the growth phase, a temporary snow fence can be used, or fast-growing shrubs can be planted between the tree barrier and the protected area and removed later. Coniferous trees are generally used, placed in rows 8 to 10 ft (2.4 to 3 m) apart with an in-row spacing of 8 ft (2.4 m). Trees are considered fully effective when their average snow trapping efficiency reaches 75 percent. Wide, dense tree plantings termed "snowbreak forests" are frequently used in Scandinavia. They collect all snow on the upwind side. A setback of about 100 ft (30 m) is recommended. Trees planted on the south side of a road should be placed so that the sun will reach the road surface at noon on the shortest day of the year.

Railroads were probably the first to use living plants for control of drifting snow shortly after the beginning of the twentieth century. Use by states did not follow until farm-to-market roads needed to be kept open during winter. Since the 1920s, many states have installed living snow fences (e.g., Colorado, Iowa, Kansas, Michigan, Minnesota, Montana, Nebraska, Pennsylvania, South Dakota, Wyoming). Three of these states have recently reported their experiences in establishing and maintaining living snow fences: Colorado (Shaw 1989), Wyoming (Powell et al. 1992), and Minnesota (Walvatne 1992). Their experience has covered long enough periods to indicate many important principles to follow to achieve success.

Multiple benefits can be obtained by use of living snow fences: in addition to snow control, they provide wildlife habitats, environmental beautification, and water resource augmentation. Research in Wyoming has shown that a fence with 50 percent porosity is most effective in trapping blowing snow. Selection of plant species to use is based on snow volume to be stored, species adaptability and longevity, soil type, soil pH and fertility, and species resistance to snow breakage. Where large amounts of drifting snow are to be trapped, two or more rows of tall, dense evergreens will be adequate. One or two rows of tall shrubs may suffice for lesser amounts of blowing snow. A rule of thumb for placement of the furthest downwind row in open terrain is no closer to the road than 200 ft. Spacing between multiple rows should be 50 to 75 ft. This requirement for sufficient right-of-way for effective snow storage is frequently an obstacle.

Survival and growth of living snow fences following planting presents few problems where climate is conducive to natural forest growth. However, watering may be necessary until the barrier is well established. Weeding, rodent control, and replacement of dead trees may also be required. Wyoming found that mortality after the first winter for a new installation was 3 percent. Rodent control was necessary by the fourth year, when mortality increased to 5 percent, but following this control, mortality declined to less than 0.5 percent.

Costs. Growth of trees and shrubs is slow, and the cost of installation must be incurred many years before the plants reach sufficient size to function as an effective snow barrier. A structural fence, of course, will be effective immediately. Wyoming has estimated the life of a structural barrier to be 35 years, whereas a living snow fence may have a useful life of 60 years. The state's estimate of 1983 costs for installation was $36,405 per mile for a living snow fence and $58,000 per mile for a 14-ft Wyoming wood fence, but over the service lives given above, the cost per mile per year would be very similar (about $1650 per mile per year). The Minnesota report has listed the following costs (1986 data) for installation of a living snow fence reported by several states:

State	Cost/mile ($)
Minnesota	50,000–200,000
Wyoming	40,000
South Dakota	32,260
Iowa	21,000
Colorado	19,000
Nebraska	10,000
Wyoming*	50,000–60,000

*Wyoming 14-ft high structural fence.

The report states that Minnesota's costs are higher because larger plant stock is used, plant spacing is tighter, continuous wood chip mulch is used, and a one-year maintenance and guarantee period is factored in.

3.5.3 Other Vegetative Barriers

Corn stubble or rows of corn left standing in a field will collect snow and serve as low fences. It is most effective to leave two strips of standing corn, each of 8 rows, separated by open space of 165 ft (50 m). The strip nearest to the road should be at a distance of about 215 ft (65 m).

4

Maintaining a Safe, Trafficable Roadway

4.1 Traction

Traction is the force developed between the driving wheels and the surface on which the vehicle is supported. In order for a vehicle to move along a road, tractive effort is required. This is provided by a force, either a pushing or a pulling force—pulling when your car is stuck or disabled and the tow truck comes to your rescue, pushing when you require a push start with a weak battery or malfunctioning starting motor. However, here we are concerned with the traction needed to propel the vehicle along its path. Traction is dependent on a force to overcome the many resistances to motion. That force is the friction between tires and pavement.

4.2 Importance of Tire/Pavement Friction

Wheeled vehicles may be able to move through dry snow that is several inches deep, but a layer of ice a fraction of an inch thick on the pavement can completely prevent movement (or perhaps allow movement in an unwanted direction, such as into a ditch or down a hill). Whether a vehicle is pushing through snow or sliding on ice, friction is necessary for traction. Friction is something we encounter every day. For example, we depend on a necessary minimum friction between our shoe soles and the surfaces on which we walk to keep from falling down. However, it is a concept that is far more complex than initially meets the eye. In order to paint a clear picture of a word that is often used loosely, some groundwork must be laid.

4.3 Definition of Friction; Coefficient of Friction

Friction is a force that opposes relative motion between two sliding surfaces. Figure 4-1 depicts such a relationship; it shows an object resting on a nonlubricated surface—a block of wood on a table.

Since the block neither falls through the table nor rises into the air, the weight of the object is exactly matched by a force exerted by the table acting in the opposite direction to the weight. This is called the normal force, designated N in the diagram, and is perpendicular to the sliding surface (it is called the "normal" force because in mathematical terms, *normal* means perpendicular). Though W is drawn on the top of the block, it actually would act on the surface. F_1 represents the force that is applied to move the block, which is resisted by the force F_2 because of the roughness between the two surfaces. In order for the block to begin moving, F_1 must be greater than F_2 to overcome what is called the static friction. Once the block begins moving, it will move at constant speed as long as F_1 equals F_2. If F_1 is greater than F_2, the block will accelerate, i.e., it will move with increasing speed. Many experiments have shown that to a high level of accuracy the force F_1 is a fraction of the normal force (weight) which depends on the nature of the contacting surfaces. This is described by the equation

$$F = \mu N \quad \text{or} \quad \mu = F/N$$

μ is the Greek letter mu and is commonly used to represent this fraction, that is, the ratio between sliding force and normal force. The

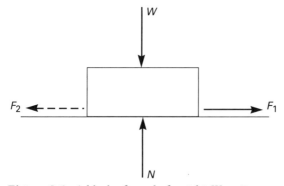

Figure 4-1. A block of wood of weight W resting on a table is supported by the normal force N. Pulling the block parallel to the table surface is resisted by the force F_2; this is the static friction until movement starts, then it becomes the sliding (kinetic) friction, which is lower than the static value.

name given to μ is *coefficient of friction*. Note that since it is a ratio of two forces, it is a pure number. This means it has no dimensions (such as feet, meters, gallons, etc.).

4.4 Factors Affecting Friction

Classical laws of friction date from the Middle Ages. Leonardo da Vinci (1452–1519) described the concepts of friction in his notebooks and included pencil sketches of several types of apparatus he designed for studying sliding friction. The first two classical laws of friction, though, are attributed to Guillaume Amontons (1699). They state:

1. The friction force is directly proportional to the applied load.
2. The friction force is independent of the apparent contact area.

These "laws" were unquestioned for many years, but we now know that beyond a fairly narrow range of loads and speeds they are not precisely true. (However, Amontons did state that friction is dependent on the materials involved, which is indeed true.) In particular, the friction between a rubber tire and the road surface is affected by a complicated set of factors which do not conform to the classical laws of friction. Some can be measured and even influenced by the driver, whereas many depend on factors beyond one's control. These influences include:

Tire factors
- Rubber composition
- Carcass construction
- Tread design
- Inflation pressure
- Size

Vehicle operating factors
- Wheel load
- Speed

Pavement surface factors
- Material composition
- Micro and macro texture

Environmental factors
- Temperature (particularly of pavement)
- Water/slush/snow depth
- Ice

4.5 Slip Ratio

Slip is defined as the difference between the vehicle and tire speeds. As a wheel is braked, the tire will begin to grip the surface, but beyond a certain frictional value it will begin to slip. The braked tire will rotate more slowly and make fewer revolutions and therefore will travel a shorter distance than a freely rotating tire. On the other hand, a tire which is accelerated (a spinning tire), will travel a greater distance than a freely rotating tire. This is explained in Fig. 4-2.

The actual distance the tire travels is d_b or d_s, but for convenience it is generalized by designating it d_a. Thus, for a slipping tire, $d_f > d_a$, and for a spinning tire, $d_f < d_a$. The slip ratio S has been introduced to express these relationships quantitatively:

$$S = (d_f - d_a)/df$$

where d_f = distance that would be traveled by a freely rolling tire
d_a = actual distance traveled by a spinning or braked tire.

The slip ratio is frequently expressed in terms of the vehicle velocity and the angular (rotational) velocity of the wheel. Both of these terms are proportional to the linear distances d_f and d_a. It is often more practicable to measure vehicle speeds and rotational speeds.

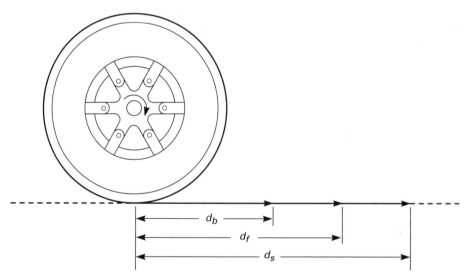

Figure 4-2. The contact surface of a freely rolling tire will cover the distance d_f. If the wheel is braked to the point that the tire slips, it will only cover the distance d_b. If, however, the wheel is accelerated and spins, the tire surface will cover the distance d_s. Though the actual distance the tire travels is d_b or d_s, for convenience it is generalized by designating it d_a. Thus for a slipping tire, $d_f > d_a$, and for a spinning tire, $d_f < d_a$.

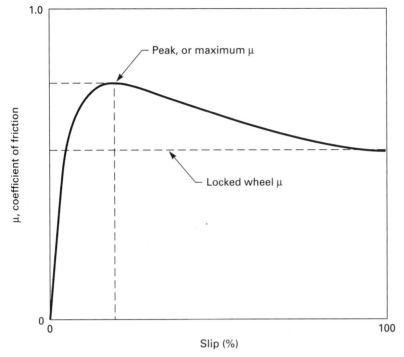

Figure 4-3. Curve describes how the friction coefficient varies with slip ratio; the peak friction varies with the nature of the surface and the tire characteristics.

Slip is important because it has been found that the value of the coefficient of friction varies with the slip ratio (see Fig. 4-3). As the slip ratio increases, the friction coefficient also increases up to a peak value, then declines steadily or may remain constant at a lower value as slip approaches 100 percent. The peak value of friction is reached at different slip ratios depending on the speed of the vehicle and the nature of the surface. Peak friction will be developed at slip ratios between about 10 and 20 percent.

4.6 Measurement of Friction

The use of friction measurements as an essential piece of information for choosing the optimum treatment strategy during a storm event has been growing in Europe, Japan, and the United States. The objective of an enlightened snow and ice control policy is not bare roads at all times, but rather a friction level on the pavement surface that does not drop below what may be considered safe for the storm condition. It should be understood that this safe level will most likely entail a reduction in vehicular speeds. The safe level will also differ for flat and hilly terrain. Sloping roads will require a higher friction value in order to obtain stopping characteristics similar to those on a level road. Thus the timing and selection of the optimum treatment should be based on measurement of the coefficient of friction. Measurements should be made often enough to reduce the number of times that the friction drops below the level considered unsafe. This practice will also reduce the number of times that pack will develop and thus will avoid the extra effort and material that deicing will most likely require.

The accurate measurement of pavement friction has been a subject of considerable research activity for many years. The complexity of the mechanism of friction has led to many theories and many attempts to explain the process satisfactorily. The fact that there is disagreement with respect to these mechanisms, and that the search for a definitive test procedure continues, need not weigh heavily on the maintenance manager who needs a practicable method for making friction measurements in order to make an informed decision regarding the best treatment strategy for the prevailing conditions. Precision is not necessary in order to provide useful guidance, although a more quantitative determination of friction is necessary for airport runways than for highways, since the value reported is a critical factor in restricting takeoffs and landings. The principles of measurement apply equally in both cases, but the selection of method and equipment must be based on the precision required.

4.7 Measurement Methods

Development of friction measurement methods and equipment was originally undertaken to assess the "skid resistance" of road surfaces. However, pavement friction is increasingly being used operationally to evaluate the performance of a snow/ice control treatment and to provide information for determining if additional treatment is necessary or if the treatment strategy should be changed. This has resulted in more intelligent use of chemicals and in many cases to a reduction in the amount used. Measurement equipment has progressed from simple stationary tests using rubber blocks, to a pendulum, to mobile equipment capable of making continuous measurements at moderate highway speeds. Practical techniques for the measurement of coefficient of friction fall into one of three groups: stopping distance, deceleration, and wheel slip.

4.7.1 Stopping Distance

This is the simplest method and requires no specialized equipment, since a passenger car or light truck can be used. The distance the vehicle skids when braked from a known speed provides the data for calculating an approximate value for average friction according to the equation (Ludema and Gujrati 1973, Hearst 1957)

$$\mu_a = V^2/30S$$

where μ_a = average friction coefficient
V = speed at instant of braking (mi/h)
S = length of skid (ft)

The friction that is measured with this locked wheel stopping distance test is not a precise value. It is referred to as "average skid performance coefficient" (SPC). Although this is a reliable method, the results can be used only for comparison purposes and not as a measure of true friction. The factor of 30 in the denominator is valid only where all wheels of the test vehicle are sliding.

4.7.2 Deceleration

We have all experienced the consequences of Newton's laws of gravity when we brake a car and then have to pick up from the floor the books and all the other loose items that had been on the seat alongside. A body in motion will stay in motion unless an external force acts on it. That is the principle of the decelerometer method of measuring friction: A pendulum will swing forward when its supporting frame is restrained. In practice, a small mass in a sensor acts on a strain gage and generates a signal proportional to the force of deceleration. A logic circuit based on Newton's second law, $F = ma$ (where F = force, m = mass of the decelerating object, and a = change in velocity), operates on this signal to give a direct readout of friction. (There's the practical application of this method in the story of the maintenance supervisor who uses what he calls the "Irish setter" friction test: If his dog slides off the seat when he makes a panic stop, he knows the road needs treatment.)

4.7.3 Wheel Slip

Mobile devices have been developed for continuous measurement of friction. Various designs utilizing a slipping wheel are used for network pavement skid resistance classification. The more sophisticated devices are also used for tire and pavement research. The slipping wheel may be installed in a trailer or in the vehicle itself. Devices using the wheel slip principle can be categorized into four groups based on the geometry of the measuring wheel and the degree of slip.

Locked Wheel Longitudinal Braking Friction. A tire following a path parallel to the vehicle travel is fixed so that it will neither rotate nor undergo any side slip. This mimics the panic stop of a car with conventional brakes. Since the same area of the tire is in contact with the abrading surface of the pavement, tire wear can produce flat spots which will affect measurement accuracy. Furthermore, as was shown in Fig. 4-3, a tire with 100 percent slip will produce a friction value that will be some indeterminate value below the peak friction.

Peak Longitudinal Braking Friction. A wheel with provision for variable slip will describe a slip-friction curve of the form given in Fig. 4-3. The degree of slip necessary to produce the peak or maximum value is maintained. Since the tire is not scuffing as much as a locked wheel, tire wear is not as great. The value measured is more representative of the pavement surface condition than that produced by a locked wheel.

Fixed-Slip Longitudinal Braking Friction. Some value of slip in the vicinity of the peak is chosen and the wheel is fixed to slip at that value. The advantage is the need for a less sophisticated mechanism.

Fixed Lateral Friction. Side-slip of a tire describes the cornering performance. It is measured by introducing a fixed angular displacement of the wheel (toe out or toe in) so that the tire is continuously slipping as it tries to follow a curving path but is constrained to move straight ahead.

4.8 Critical Values of Pavement Friction

Any contaminant on a pavement surface will reduce the coefficient below the level under clean, dry conditions. Table 4-1 lists the range of coefficients of friction during braking (locked wheel) found in tests in Japan (Ichihara and Mizoguchi 1970).

Table 4-1. Friction Coefficients during Braking

Snow or ice condition	Friction coefficient
Ice	0.1–0.2
New snow	0.2–0.25
Old snow	0.25–0.30
Refrozen snow	0.30–0.40
Chloride-treated snow	0.35–0.45
Sand-treated snow	0.30–0.40
Chloride-sand mixture	0.30–0.50

4.9 How to Increase Friction Values

Tire friction will be reduced by a thin water film that will always be present to a degree dependent on the air temperature, relative humidity, pavement type, and concentration of salt in the film. The thickness of the liquid film will vary from 0.004 in (0.1 mm) for rain-wetted roads to 0.00000004 in (1 nm) on clean, dry roads (Mortimer and Ludema 1972). Chemicals applied to the surface have a very significant effect on evaporation rate of the liquid film and therefore on the time and rate of transition of tire friction from low to high values. Chemicals reduce tire friction below the plain-water level and prolong the reduction by lowering the evaporation rate as a consequence of the reduced bulk water vapor pressure. Because calcium chloride is deliquescent (absorbs moisture from the air and goes into solution), it will retain a thin liquid layer at humidities above about 30 percent, whereas films containing dissolved sodium chloride and most other ice control chemicals will become dry and the chemicals will crystallize. In general, the higher the chemical concentration in the film, the more slowly friction will increase, and the higher the humidity of the air, the longer chemical contaminants will prolong the transition from wet to dry friction values. Measurements have also shown that the reduction in friction on calcium chloride–treated pavement can be twice that caused by sodium chloride.

Bits of abraded tire rubber and grease drippings from vehicles will always be present on in-service pavements and will lower the friction coefficient, particularly when wet. The first rain following a dry period will cause a sharp drop in the friction value until some of the contaminant washes away. Maintenance personnel do not attempt to improve the frictional characteristics of pavements in these conditions. However, ice or snow can cause much greater reductions in friction with a longer-lasting effect and will require some remedial treatment. Three treatments are available: Mechanical removal of the ice or snow to expose as much of the bare pavement as possible, application of chemicals to melt the snow or ice to expose the pavement, and application of abrasives to provide a temporary increase in friction.

4.10 Abrasives for Friction Improvement

Sand and other gritty materials have a long history of use for treating slippery pavements to improve friction characteristics. Until the middle of the twentieth century, salt or other chemicals were not widely used. The common practice in that era of waiting until 1 or 2 in (2.5 to 5 cm) of snow had accumulated on a road before commencing any control activity resulted in the development of "pack," a compacted snow mat tightly bonded to the pavement. Removal was difficult and slow, so an antiskid material was applied for traction improvement. The amount of salt required for removal of this pack began to climb about 1955 as the motoring public increased its demands for more rapid friction improvement. Many communities and states implemented "bare road" policies, which generally depended on application of salt for success. By the early 1970s, however, the unforeseen consequences of salt intrusion into water supplies, damage to vegetation, corrosion of vehicles, and deterioration of bridge decks and other parts of the highway infrastructure became more and more apparent (Anon 1971). This led to increased use of antiskid materials in an attempt to ameliorate the problem. Here also, unforeseen consequences of the strategy began to appear, with increasing problems of clogging of drainage channels and sewers, costly cleanup of piles of traffic-blown leftover abrasive material, and most recently the pollution resulting from airborne fine particles—the PM_{10} problem, which will be elaborated on in Sec. 4.10.3. Recent research has demonstrated that application of abrasives is unnecessary on well-maintained, heavily traveled roads where timely applications of an ice control chemical as an anti-icing treatment have been made, and in fact in some instances use of antiskid material has reduced tire-friction values. (See Sec. 4.12.1 for a discussion of anti-icing.) In any event, antiskid materials such as sand will not melt snow or ice, nor will they assist in breaking the ice-pavement bond; their only function is to provide temporary friction improvement.

4.10.1 Materials and Their Effectiveness

Material for antiskid applications should meet these requirements:

- It should be insoluble, or at least only slightly soluble, so that it will not dissolve completely—if it dissolves, it will no longer provide a gritty surface.
- The particles should be hard and no smaller than coarse sand nor larger than small pebbles [about ¼ in (6 mm) in its largest dimension] to cover the treated surface most economically and thoroughly.

  Use of ashes or anything that has dirt in it is a poor idea. They tend to make an icy surface even more slippery, since particle size is very small; they are soft and generally at least partially soluble; and they are so light that a slight breeze may blow them off the ice unless the surface is wet. Silt-size particles (particles less than 75 μm in diameter) contribute to the PM_{10} loading (see Sec. 4.10.3).

- The best particles to use are those with sharp edges, the better to dig into the ice and provide increased friction. Concrete (or "manufactured") sand is better than beach sand because the particles are sharp rather than rounded. Manufactured sand is produced by crushing hard rocks and screening out the large pieces. This leaves the small, sharp-edged particles, which are most effective for ice treatment. (This type of sand is sometimes called "concrete" sand, since its use results in higher-quality concrete.) Quartz makes the best sand, although hard limestone is also suitable. Soft rocks such as shale and sandstone are undesirable. Cinders or clinker may be available in some parts of the country, but they frequently have residues of sulfates, heavy metals, and other environmental bad actors. If they are to be used in any quantity, their composition should be analyzed.

The Environmental Protection Agency has determined the desirable properties of antiskid materials to reduce creation of fine particles; see Table 4-2.

There is no "standard" or optimum gradation of sand for antiskid applications. Alaska highway personnel evaluated several gradations and found the following the most effective in field trials (Conner and Gaffi 1982):

Sieve size	Maximum percent passing
3/8	100
1/4	98–100
#16	50–90
#100	15
#200	0–7

Table 4-2. Criteria for Selecting Antiskid Materials to Reduce Fine Particle Generation

Measurement parameter	Units	Mean value for acceptable materials	Mean value for unacceptable materials
Modified Los Angeles abrasion loss	Weight %	3	11
Initial silt content	Weight %	0.1	6
Vickers hardness	kg/mm^2	1000	800
Particle shape index	Dimensionless	10	9

Note: If the properties fall between acceptable and unacceptable, a material is considered questionable and "good engineering judgment should be used."

Source: Kinsey et al. (1990).

Table 4-3. Damage to Glass Windshields from Flying Antiskid Particles

Particle size	Thrown by auto going 50 mi/h (80 km/h)	Thrown by auto going 30 mi/h (48 km/h)
#4–#10 (4.75–2.0 mm)	Heavy pitting	Heavy pitting
⅜ in–#4 (9.5–4.75 mm)	Pitting and stars	Stars, Types 1, 3 cracks
½–⅜ in (12.5–9.5 mm)	Types 1, 2 cracks	Types 1, 2, 3 cracks
¾–½ in (19.0–12.5 mm)	Types 2, 3 cracks	Types 1, 2, 3 cracks

Pitting: small superficial chips; stars: similar to damage caused by pellet gun; type 1 crack: ½ to 1 in at the base; type 2 crack: series of concentric arcs surrounding impact point; type 3 crack: series of radial cracks extending from impact point with no cratering.

Source: Conner and Gaffi (1982).

The Alaska tests also investigated the breakage of windshields by various sizes of particles. It was found that spherical particles caused more damage than rough-edged fractured particles. Several factors account for this behavior: Rough, angular particles create more wind resistance and slow down, they break more readily and also rotate on impact, and they may have less mass than spherical particles for the same screen fraction. All these behaviors absorb energy and provide additional reasons why crushed material is preferable. The tests also concluded that particle size should not be greater than $\frac{1}{4}$ in (6 mm) to avoid windshield damage. The extent of the damage reported in the tests is given in Table 4-3.

4.10.2 Application of Material

Dry antiskid material will readily blow off a cold ice- or compacted snow–covered surface. It has become common practice to add a freezing-point depressant chemical to the material to assist in wetting the surface for better adhesion. A chemical also serves to prevent formation of frozen lumps which can interfere with spreading. Addition of 5 to 10 wt% of chemical is most common. Salt is added when temperatures are not expected to be below about 20°F (-7°C) at time of application, and calcium chloride is substituted either partially or entirely at lower temperatures. Application rates that agencies report using range from 100 to 1200 lb/la-mi (8 to 92 g/m^2). No comprehensive tests have determined the optimum amount of material to apply for specific conditions.

4.10.3 Effect of Abrasives on Air Quality

The United States has established as the national primary and secondary 24-h ambient air quality standard for particulate matter a limit of 150 $\mu g/m^3$ for a 24-h average concentration. There is also an annual arithmetic mean standard of 50 $\mu g/m^3$. Particulate matter is measured in the ambient air as PM_{10} (particles with an aerodynamic diameter less than or equal to a nominal 10 μm) (Code of Federal Regulations). One source of PM_{10} emissions is the resuspension of antiskid materials into the air; the constant grinding by vehicle tires will abrade the hardest antiskid material into fine particles. Many areas of the country exceed these limits, and as a consequence have had to take steps to reduce their use of abrasives for ice control. Environmental Protection Agency studies indicate that the salt mixed with abrasive material applied for antiskid purposes contributes an insignificant amount of particulates to PM_{10} loading. Instead, salt aids in clearing the road by forming slush that is either carried away on the underside of vehicles, cast off the road, or removed as runoff (Kinsey 1995).

4.10.4 Other Problems; Cleanup

A major difficulty with application of large amounts of antiskid materials arose in many jurisdictions even before the fine dust problem made sweeping up the residue a necessity. Surface drainage channels were becoming blocked by the accumulation of solid fines, and sewer capacities were being reduced as well. Sweeping the residue is a costly and time-consuming activity; one city estimates that it requires three to four weeks to clean up its streets after only one day of abrasive spreading. Communities with combined sewers (those handling both industrial and residential waste and storm runoff) have found that the large quantities of particulate material affect aerobic digestion in the sewage treatment facility, and there is the added problem of sediment removal and disposal. A further problem results from the reduced distance a truckload of abrasive will cover compared to a load of salt or other chemical. The additional time spent returning to the supply yard to obtain another load of abrasive delays the completion of treatment, may extend the workday and lead to the added cost of overtime, and adds extra mileage on the truck.

4.10.5 Storage of Materials

Sand or other abrasive must be kept dry if it is stored at below-freezing temperatures, as any moisture present may form hard lumps. Mixing salt in the sand as it is stockpiled is the most frequently practiced method. Liquid calcium chloride is also used. Section 3.2.7 includes information pertinent to the storage of abrasives.

4.11 Mechanical Methods

Additional methods for improving the friction characteristics of snow- or ice-covered pavements are available. The most fundamental method is the removal of as much snow as possible by mechanical means (see Figs. 4-4 and 4-5. This is advisable under almost any conditions, the exceptions being when air and pavement temperatures are cold, the snow has low water content, and there is enough traffic or wind to scour the road. It is especially important to remove snow before traffic can pound it into a compacted slab that may be tightly bonded to the pavement. Complete removal of snow using blade plows is not usually possible because there are irregularities in the road surface, the plow may bounce, or there is spill or blowback. And, of course, clearing during a snowfall will leave snow accumulating on the road that has just been plowed. There is no residual effect from plowing. For that, chemical treatments are necessary.

4.12 Chemical Methods

Although many methods for maintaining roads with a coefficient of friction at a level acceptable for the type of traffic on the road have been tried, chemicals have assumed a major role because of their effective performance, relatively low cost (in comparison to alternatives), and residual or carryover effect. Salt (sodium chloride) is such an

Figure 4-4. Typical one-way displacement plow clearing deep snow.

Figure 4-5. Use of wing plow for widening and "benching" a windrow (unit is a grader equipped with front V-plow and center blade as well as wing).

effective and low-cost material (though not without environmental problems, which must be included in the total cost of its use) that, it has been said, if such a material did not exist, it would have to be invented. Because of its low material cost, it has in the past been abused, and this has led to efforts to apply it smarter—which means at a lower application rate. Improved application equipment, described in Chap. 5, have helped in this respect, but a recent fundamental change in treatment philosophy has the promise of paying the biggest dividends. The technique offering this promise is anti-icing.

4.12.1 Anti-icing

Anti-icing is the "practice of preventing the formation or development of bonded snow and ice by timely applications of a chemical freezing point depressant" (Ketcham et al. 1996) (see Figs. 4-6 and 4-7). In truth, it is not a new technique. It has been practiced to some extent for many years by forward-thinking maintenance managers, with varying degrees of success. It is the recent development and availability of new technology essential for its success that has made the difference. Successful anti-icing depends on good timing—on the ability to apply a relatively small amount of chemical to a pavement before frozen or freezing precipitation can bond to the pavement. Doing so will inhibit the formation of a bond; late treatment will most likely result in the development of "pack," the compacted snow layer tightly bonded to the pavement that is extremely

Figure 4-6. Applying liquid ice control chemical with towed unit; note the side nozzle spraying the center of the lane.

Figure 4-7. Liquid chemical storage facility with spray boom for truckload prewetting.

difficult to remove by plowing and requires deicing for removal (see the next section for coverage of deicing, a technique that may be acceptable in some circumstances). Two pieces of information are required to make an informed decision about when to start a chemical treatment: the pavement surface state and a reliable weather forecast. Pavement surface temperature and its trend will indicate whether there is a high probability of snow or ice freezing on the pavement, and the weather forecast will indicate the likelihood of the arrival of precipitation and its most likely form. The advent of road weather information systems (RWIS) now provides the means for assessing the pavement surface condition in real time. The use of satellite imagery, Doppler radar, a dense network of weather reporting stations, and supercomputers capable of rapidly modeling the dynamic atmospheric forces all have led to better, more reliable forecasts. In addition, the skill of private meteorological services to augment the large-scale weather forecasts and detailed data available from government sources can increase the reliability of meso- and microscale forecasts. The subspecialty of road weather meteorologist has developed, with demonstrated benefits to the highway community.

The Federal Highway Administration (FHWA) has published a detailed guide for anti-icing: *Manual of Practice for an Effective Anti-icing Program: A Guide for Highway Winter Maintenance Personnel* (Ketcham et al. 1996). A print version is available from FHWA, or it can be downloaded from the FHWA's World Wide Web site at http://www.fhwa.dot.gov/reports/mopeap/eapcov.htm. The ready availability of this report makes it unnecessary to provide details extracted from it, and it is recommended as a comprehensive source of information concerning contemporary effective practices.

4.12.2 Deicing

Deicing is necessary when ice or compacted snow is strongly bonded to the pavement and the bond has to be destroyed in order to remove the frozen layer. Deicing requires heavier applications of chemical than anti-icing as a result of dilution of the particles as they bore through the layer to reach the ice-pavement interface. Customary application rates are 250 to 500 lb/la-mi (19 to 38 g/m^2), depending on temperature and amount of snow or ice to be removed.

4.13 Thermal Methods

A third way of improving friction of ice- or snow-covered roads entails heating the pavement surface to prevent the formation of a bonded slippery layer, or to melt such a layer if one has formed. Further details of the technology available for bridge and road heating can be found in Sec. 3.4.

5
Snow and Ice Control Equipment

5.1 Introduction

The response to frozen precipitation on a road will depend on the nature of the precipitation. If it is snow of depth sufficient to impede motorized travel, plowing is the appropriate response. If it is ice or a thin snow layer, chemical treatment may be the best method. In either case, equipment specific to the task should be utilized. This falls into two major classes: snow removal equipment and ice control/removal equipment.

5.2 Snow Removal Equipment

Wheeled vehicles are not normally able to travel over level snow-covered pavements when the depth exceeds half the wheel diameter. Modern economies require unimpeded year-round mobility to avoid serious social and economic dislocations, and this, together with the increased probability of accidents on snow- and ice-covered roads, dictates the removal of snow from the traveled way. Clearance of snow from airport runways, taxiways, aprons, and ramps is also necessary to ensure safe operations. Preventing the development of slush is of greater concern at airports than on highways because of the high landing speeds of aircraft and the danger of hydroplaning. Railroad tracks must be cleared to reduce the frontal resistance to locomotives as well as to prevent derailments caused by the wheel flanges riding up on the compacted snow. All major transportation systems have a common need to reduce snow accumulation to a level which allows safe passage. This chapter describes the characteristics of snow and ice removal equipment with particular reference to highways, though many pieces of highway equipment find use on railroads and airports. Equipment developed for the unique problems of railroad snow removal is described in Chap. 10. With only a few exceptions, airports use equipment little different from that used on highways; airport snow and ice control is described in greater detail in Chap. 11.

Snow falling on a paved surface may be removed by chemical, thermal, or mechanical means. Chemical and thermal methods are discussed in Chap. 3. Mobile mechanical equipment, including devices for removing ice, is described in this chapter. Frequently the same equipment is used to remove both snow and ice. However, because of the high-strength adhesive bonds which may form between ice or compacted snow and pavement, specialized equipment is frequently required for satisfactory results.

5.2.1 Evolution of Equipment

The snowplows we see today on highways and airports owe their basic design to the railroads. The history of this development is presented in Chap. 10. There is early mention of the use of horse-drawn plows to clear railroad tracks in 1831 and to clear city streets in 1862. U.S. patents for horse-drawn rotary sweepers were issued to three New York residents in the early 1860s. However, horse-drawn sleighs were the primary method of conveyance both in cities and in rural areas, and clearance of snow was resisted since it interfered with their operation. Instead, snow rollers were used to compact the snow for better sleighing. The transition to highway snow removal devices commenced with the expansion of street railways in the major eastern cities in the United States in the last decade of the nineteenth century. In 1905 the trolley companies in Boston and Worcester, Massachusetts, began using a newly designed rotary plow, one mounted on each end of heavy trolley cars.

At the time the first American automobile, the Duryea, was built in 1892, the Good Roads Movement in the United States, inaugurated at the behest of bicyclists in the League of American Wheelmen, had begun the upgrade of the muddy tracks called roads. Rural free delivery of the United States mail was inaugurated in 1896. By 1900, seven state highway departments had been created. In 1902 both the American Road Builders' Association and the American Automobile Association were formed. A big impetus to highway construction came from the Federal-Aid Road Act of 1916, which provided federal funding for construction (Anon 1971a). World War I greatly increased the production and use of trucks. But although the road network was growing rapidly, it largely consisted of disconnected pieces, and road travel outside of cities in winter was difficult and often impossible. Little effort was made to clear snow from rural roads. This piecemeal development of highways was replaced by an integrated road network as a result of legislation passed in 1921 requiring that a federal-aid system be designated and that all federal assistance be concentrated on it. The need for removal of snow from roads increased, and the number and types of truck-mounted plows increased commensurately (Minsk 1970).

5.2.2 Plows

The design of any plow is a compromise, as was discussed in some detail in Sec. 2.5.3, and we have seen that in the process of pushing into the snow, lifting it, and casting it sufficiently far from the cleared area to reduce the necessity for rehandling it, compaction of snow absorbs a high percentage of the truck power. It is the objective of minimizing this energy loss and smoothing the flow of minimally disturbed snow that has driven the design of plows. However, the three functions of a displacement plow cannot all be accomplished with equal efficiency by any one design. That is why all plows are compromises, and a design may succeed in maximizing one, or perhaps two, of these functions, but rarely all three.

Displacement Plows. This type of plow uses a concave moldboard to move, or displace, the snow from the area being cleared. It is generally referred to as a blade plow. This is the most common type of plow, and it is most frequently mounted on the front of the vehicle. Its function is to cut the snow from the road, lift it, and change the direction of flow from the direction of the truck movement to a direction approximately at a right angle so as to cast the snow to the side, away from the cleared area. Blade plows are also mounted in other locations on the truck to accomplish additional functions as described below. A "family tree" of types of displacement plows is given in Fig. 5-1.

Front (or Nose) Plow. As the name implies, this type of plow is mounted on the front of the truck. Manufacturers offer many design combinations of widths, heights, and shapes of plows. A plow is typically angled at 35° from perpendicular to the plowing direction, a position that experience has demonstrated is effective for collecting snow and casting it to the side. Figure 5-2 identifies the components of a typical

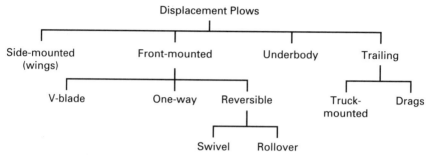

Figure 5-1. Types of displacement plows.

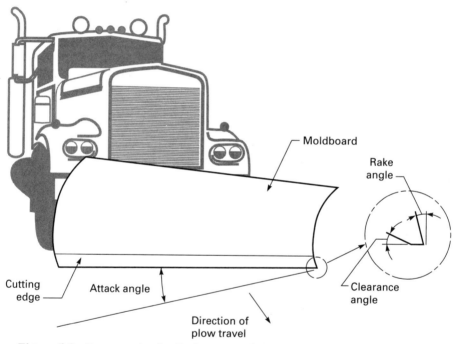

Figure 5-2. Components of a displacement plow using common terminology.

displacement plow. Displacement plows have three types: one-way, reversible, and V.

One-way plows can change neither their shape nor their orientation but are designed always to cast the snow in the same direction, either to the right or to the left, or sometimes in the "bulldozer" position, i.e., perpendicular to the direction of movement (Fig. 5-3). The fixed cast

Snow and Ice Control Equipment

Figure 5-3. Front-mounted one-way snowplow and wing on a medium size (34,000-lb) GVW truck.

direction enables the design to be optimized for the three functions described above, but it also requires the truck to remain on one side of a lane to avoid compromising safety. Plowing with a train of one-way plows (in echelon) on a multilane highway can be done safely, however.

Three types of reversible plows are commercially available: the rollover, the swivel, and the deformable moldboard.

- The simplest and most common design of reversible plow is the swivel. Its blade is symmetrical about a vertical plane normal to the blade. It is shifted from one cast direction to the other by a crank, pivot, or eccentric linkage rotating about a vertical axis. Modern reversible plows of this type generally use hydraulic actuators to accomplish this shift. Manually shifted plows are still used in many situations where a change in cast direction while the truck is under way is not required; they offer the additional virtue of less mechanical complexity and lower cost.

- The rollover rotates about a horizontal axis parallel to the direction of vehicle movement and has a conical-shaped blade that is symmetrical about a plane parallel to the pavement (Fig. 5-4). This type of plow is more costly than simpler reversible plows. However, there are additional reasons why its use has been limited to airports:

Figure 5-4. "Rollover" front-mounted plow in process of rotating its cast direction.

It is heavy and requires a large truck for mounting, and the mount extends to the front of the truck to such an extent that a wide turning radius is required. This is no problem on airport runways. Its advantage lies in the increased efficiency of snow handling, since the shape can be optimized for both right and left casting, and so a smoother flow of snow can be maintained.

- A deformable moldboard can be changed in shape and position to form either a left- or right-cast one-way plow. This reversibility is achieved by changing the curvature using hydraulic rams controlled from inside the cab. A typical unit is made of $3/8$-in-thick polyethylene sheet, which also provides a noncorrosive, low-friction surface that reduces resistance to the flow of snow.

The V-blade plow is a variant of the fixed-blade plow that can cast snow to the right and left simultaneously. Some designs can be shifted either to the left or to the right to a small extent to favor casting in one direction, but they do not achieve the casting efficiency of the other designs. V-blade plows are usually large, heavy units mounted on trucks, though smaller types are sometimes mounted on front-end loaders and graders. They are used primarily for "pioneering," i.e., making the first pass down a road closed by heavy snow to open it

Snow and Ice Control Equipment

and allow the entry of the lighter but faster one-way or reversible plows.

Wing (Side-mounted) Plows. Front-mounted plows clear a width not much greater than the truck width. Blades that can be extended to one or both sides of the truck can increase the plow swath so that the plow can clear an entire lane or a partial lane and the shoulder in a single pass. These wing plows are generally cylindrical in cross section, since their primary function is to channel the snow to the side. They can be mounted in a number of ways. The common method in the United States uses a kingpost located either on the front of the plow hitch or immediately in front of the cab. Outside North America, a common attachment position is midway between the front and rear axles or behind the rear axle.

Underbody Plow. Underbody blades mounted underneath trucks (Fig. 5-5) are similar in function to the blade on a grader (motor patrol). A grader is frequently used when ice or compacted snow has collected on a road and mechanical removal is desired. The capability of applying downpressure on the blade by means of hydraulic actuators helps to penetrate or scrape the ice. Trucks equipped with underbody blades are capable of operating at much higher speeds than graders and are used by some agencies for plowing new snow as well as for removal of ice and compacted snow. The use of underbody blades also improves the maneuverability of the plow truck by reducing the overhang of a front-mounted plow.

Figure 5-5. Underbody blade cutting ice from pavement.

There are three major disadvantages to the location of this plow: (1) The front wheels of the truck carrier precede the plow and therefore may compact the snow and make its removal by the blade difficult, (2) the front, steering wheels are operating on uncleared pavement, in contrast to those of a front-mounted plow, and (3) the blade is not visible by the operator, and as a consequence the outer edge of the blade often catches on an obstacle such as a curb.

Trailing Plow. A plow mounted at the rear of the truck carrier is called a trailing plow. This design has the same disadvantages as the underbody plow. It is seldom used today.

Rotary Plows ("Snowblowers"). Blade plows have limited casting range and are not capable of displacing very deep or very hard snow. This has led to the development of rotating cutting devices with one or more rotating elements. All designs of rotary plows cut the snow with a rotating element and also cast the snow from the cut by means of a rotating device. The family tree of rotary plows is given in Fig. 5-6.

Single-element rotary plows use the same rotating device to both disaggregate the snow and cast it from the cut. Two designs are illustrated in Fig. 5-7: the scoop wheel, whose axis of rotation is parallel to the direction of vehicle movement (axial rotation, Fig. 5-7a); and the milling drum, in which the axis of rotation is horizontal and normal to the direction of vehicle movement (transverse rotation, Fig. 5-7b).

Two-element plows have separate components for disaggregating and casting (see Fig. 5-8). Disaggregating elements rotate in the transverse direction, while impellers rotate axially in the direction of motion. Some disaggregators utilize a pair of augers (Fig. 5-8a) or a helical ribbon cutter (Fig. 5-8b). Impellers consist of a number of flat or contoured blades attached to a web or disk (see Fig. 5-9a and b, respectively).

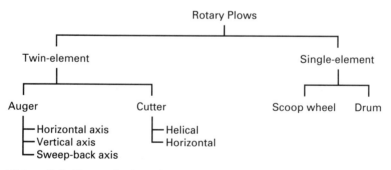

Figure 5-6. Types of rotary plows.

Snow and Ice Control Equipment

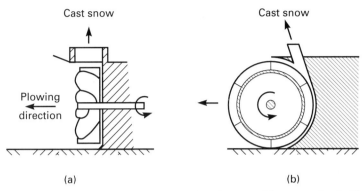

Figure 5-7. Single-element rotary plow designs: (a) scoop wheel; (b) milling drum.

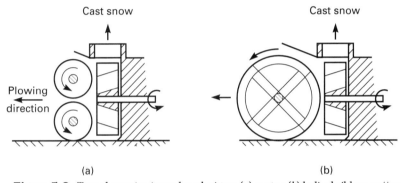

Figure 5-8. Two-element rotary plow designs: (a) auger; (b) helical ribbon cutter.

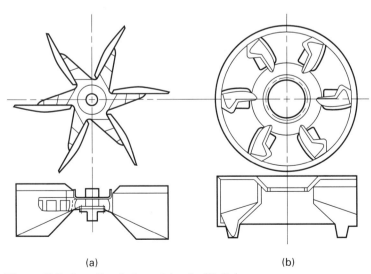

Figure 5-9. Impeller designs: (a) web; (b) disk.

Figure 5-10. Power broom with rear air-blast duct cleaning airfield runway.

5.2.3 Power Brooms

Brooms (Fig. 5-10) are generally used for snow removal on airfield runways. Since little directional control is possible, the snow that is swept up is usually cast in the direction of the prevailing wind. The bristles are made of steel or fiber (natural or synthetic). Loose snow is readily removed with a broom, but ice or well-bonded snow will not be removed completely without prior chemical treatment or softening by solar radiation.

5.3 Ice Removal/Control Equipment

Blade plows are also used for scraping and cutting compacted snow and ice in the attempt to remove them. However, the irregularities of pavements and the wear of cutting edges conspire to prevent complete removal (see Sec. 5.3.3 for more on ice-cutting blades). Chemicals for removal of ice or pack (deicing) or for prevention of ice–pavement bond formation (anti-icing) are applied using spreading equipment. Abrasives are also applied using similar equipment.

5.3.1 Spreaders for Solid Materials

The first spreaders used for applying solid ice control materials were modified agricultural limestone spreaders. This is the equipment that

Snow and Ice Control Equipment

most people think of when spreaders are mentioned. However, liquid chemicals are now being increasingly used, and though they are applied by spray apparatus, in popular terms these units are also called spreaders.

Safety is enhanced by eliminating variations in the surface condition so that unanticipated changes do not lead to loss of vehicle control. This objective is best met by applying the materials uniformly across the treatment area at a constant rate regardless of the speed of the spreading vehicle. This involves confining the distribution to a chosen, constant width at a feed rate dependent on the truck speed. The simplest equipment uses a constant feed rate to the distributor; thus the amount applied on a unit area can be varied only by changing the speed of the truck. This non-speed-related method is referred to as the manual method, since the distribution is set for a given road speed.

Conventional distribution devices use a spinner or auger (Fig. 5-11). The chemicals or abrasives are fed to the distribution device by gravity by raising the dump box, or by a feed mechanism such as a moving belt. (Zero-velocity spreaders are relatively new devices which attempt to reduce material loss by applying material with no relative velocity difference between truck and road. This equipment will be described below.) Application of materials at a constant areal coverage regardless of truck speed is the purpose of spread-rate controllers, used in so-called ground-oriented spreaders.

Figure 5-11. Spreader with spinner on left side of truck for centerline application (U.S.).

Dump Box. The simplest type of spreader uses gravity to feed the material to the distributing mechanism. In the conventional design, the forward end of the dump box lifts so that material is dropped onto a rear-located distribution device. In an alternative design, the rear of the dump box is raised to feed material to the distribution device (a spinner, as discussed later) in front of the rear wheels. This design has the advantage of placing the friction-improving material immediately in front of the driving wheels, a decided advantage when ice or compacted snow have reduced the friction to a very low level, and the spreader vehicle must get through.

Hopper Body. This is sometimes referred to as a V-box, because in cross section it has that shape. At the apex of the V is a moving belt or ladder chain with paddles to transport the material from the hopper to the distribution device. A hydraulic motor drives the belt or chain; its speed can be varied by a controller in ground-oriented units.

Distributors: Spinners/Rollers. A roller distributor running the width of the truck will apply material to that width and no more. Paddles, or projections, on the roller drag a specific amount of material to the discharge opening running the full width of the unit. The discharge quantity is varied by varying the rotational speed of the roller.

A spinner is a rotating disk placed below the truck bed, generally very close to the pavement to reduce the bouncing of material as it strikes the pavement (Fig. 5-11). Use of a spinner provides much greater control of application variables: Width of spread can be varied, as can the symmetry of the spread pattern. The shape of the vanes on the disk will influence the pattern, as will the plane of rotation with respect to the horizontal. Spreaders for airports, where a wide swath is desired on each pass, use two independent spinners, each of which can be angled to produce asymmetrical spread patterns.

Controllers for Spreaders. Hydraulic devices are used for power transfer in conventional truck equipment. That enables close control of the speed of motors and actuators used in spreaders. The vehicle speed is sensed and fed to the controller, which in turn modulates the speed of the material feed mechanism to deliver a calibrated amount of material to the distributor. The controller itself is located in the truck cab, where the operator can select the amount of material to be applied. Two types of these so-called ground-oriented automatic spreader controllers have been developed, open-loop and closed-loop (Fitzpatrick and Law 1978).

- *Open-loop (Fig. 5-12a).* This type monitors truck speed and adjusts the control valve to a predetermined setting to provide the correct belt speed for the desired spread rate. Any changes in the hydraulic system variables will result in an error in the belt speed.

Snow and Ice Control Equipment

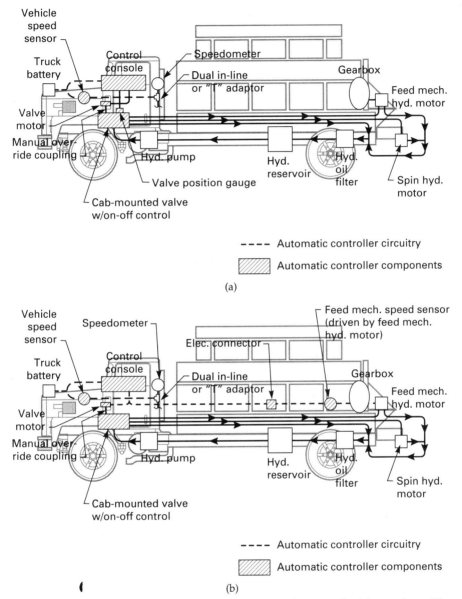

Figure 5-12. Schematic drawing of hydraulic spreader controls: (a) open-loop; (b) closed-loop (Besselievre 1976).

- *Closed-loop (Fig. 5-12b)*. This type monitors both truck speed and belt speed and adjusts the control valve until a predetermined value of the ratio of the belt speed V_s and truck speed V_t is obtained. This will greatly reduce the likelihood of a systematic error in delivery rate. Another design, pioneered in Europe, uses a separate wheel to drive the material feed mechanism in concert with the truck speed. The spinner speed may be varied independently to control the spread width.

Four advantages are cited for automatic control systems: (1) Only one variable need be set manually at the start of a spreading operation, (2) operating conditions are continuously monitored, (3) continuously variable rather than discrete adjustment of the control valve is provided, and (4) safety is increased by eliminating driver intervention while simultaneously reducing such intervention as a source of error.

Spreader Calibration. The most sophisticated controller will be of little value if an unknown amount of material is being spread. The rate at which material is delivered to the distributor depends on the area of the gate opening, the speed of the feed belt, and the truck speed. All mechanical equipment is subject to changes resulting from wear and other causes. Equipment must be calibrated periodically to ensure that the desired amount of material is actually being applied. Two methods are generally used: (1) connecting a speed simulator to the control panel so that the controller senses that the truck is traveling at a specified road speed, and (2) jacking up the rear wheels so that the truck can be run in place at a specified speed and a timed volume (or weight) of discharge measured. A less accurate method involves operating the truck over a measured distance at a known constant speed and determining the weight of material spread by measuring the load weight before and after spreading. The variation in operating parameters over a range of speeds makes it necessary to check calibration over a number of truck speeds.

Zero-Velocity Spreaders. This design for solid chemical application equipment has been developed to confine the material within a traffic lane and thereby reduce loss from material bouncing off the pavement. This enables the same treatment level to be achieved with smaller quantities of chemical and helps reduce the amount of chemical that may go into the environment. The zero-velocity design receives its name from a description of the speed of the chemical discharged from the spreader with respect to the pavement. The chemical is propelled at a speed matching the speed of the vehicle, and as a consequence most of it will drop straight down onto the road in a narrow band. Some differences in particle speed will occur because of size and mass variability, and air turbulence will affect the pattern as well, but most of the material will drop straight down. The spread pattern of spinner-type spreaders is affected adversely by vehicle speed: The faster this is, the greater the likelihood of particle bouncing and hence the greater the loss. It is often desirable to spread at traffic speeds to minimize the speed difference between the truck and other vehicles. The zero-velocity spread pattern is much less affected by truck speed, and so operation at moderate traffic speeds is an advantage. The narrow spread width is particularly advantageous for deicing, where rapid penetration of the ice and subsequent brine flow at the ice–pavement interface is desired. The narrow spread is less desirable in two situations where uniform coverage of most of a lane width is the objective: application of abrasives and anti-icing applications. The zero-velocity spreader pattern would have to depend on traffic for distribution of an anti-icing chemical over that width.

5.3.2 Applicators (Spreaders) for Liquid Chemicals

Liquid chemicals are frequently the material of choice for anti-icing treatments. Application equipment for these chemicals has progressed from home-built tank and spray nozzle arrangements with crude flow rate controls, to agricultural sprayers, to the sophisticated units now available. One type is a tow-behind two-wheel unit on which a spray bar with several nozzles is mounted (Fig. 5-13). One of the wheels drives a pump at a rate determined by the travel speed.

Figure 5-13. Tow-behind applicator for applying liquid chemicals.

5.3.3 Ice-cutting Blades

The ability of the cutting edges of front-mounted plows to shear ice and compacted snow from the pavement is very limited, since the primary function of the blade and moldboard is to lift and discharge snow as rapidly and efficiently as possible. The angle of the cutting edge must be selected to minimize "chatter" which will lead to blade bounce and possibly blade tripping. Research has led to increased knowledge of ice-cutting dynamics and improved cutting-edge design (Nixon 1993).

Carbon steel cutting edges may last no more than a few hours of plowing. Tungsten carbide inserts along the contact edge will last for many hundreds of hours of plowing. These estimates are based on the cutting edge contacting the pavement, a necessary condition if the objective is as complete removal of snow and ice as possible. An older practice, still used by some agencies, is to use caster wheels or shoes (short ski-like supports) to hold the cutting edge a short distance above the pavement to eliminate wear. This practice has declined because of the almost invariable need to use a chemical treatment on the residual snow. Contact of a steel cutting edge with the pavement, however, will peel off lane markings and some raised reflectorized buttons. This is avoided by using elastomeric cutting edges, commonly called "rubber" edges although urethane is more commonly used for the purpose.

5.3.4 Underbody Blades
(See Fig. 5-5)

Graders (also called motor patrols) are a familiar sight in rural areas, where they are used to grade gravel roads. The grader blade is located on a turntable approximately midway between the front and rear wheels. The same concept is used on truck-mounted underbody blades. Their advantage lies in the capability for controlling the amount of downpressure on the cutting edge, and thus their value in cutting ice from the pavement. The tilt angle and attack angle on most units can be varied to suit the prevailing conditions.

5.4 Snow and Ice Control Vehicles

Motorized carriers for snow removal equipment range from conventional dump trucks to specialized wheeled or tracked vehicles designed specifically for the task of clearing snow or removing ice. V-blades used for opening a heavily drifted road are most often mounted on heavy four-wheel-drive trucks, but tracked vehicles (a "bulldozer" without its straight front blade) are also used because of their greater traction. High-capacity rotary plows (capable of moving up to 10,000 tons/h) require heavy vehicles to carry both the rotary head and its high-horsepower drive engine as well as the vehicle engine. These are specialized vehicles, built for the single purpose of snow removal. Smaller rotary plow units consisting of the rotary head and its drive engine can be mounted temporarily on trucks or front-end loaders. A carrier used for this application must have adequate front-axle load rating for the heavy load the plow adds.

5.4.1 Mechanical Requirements

Plowing snow induces large stresses that are very different from those for which a standard truck is designed. Heavy loads hauled in the dump box are carried largely by the rear axle or axles. A front-mounted plow, in contrast, applies heavy unbalanced forces which may cause racking (twisting) of frames that are not designed to withstand them. Other aspects of good truck design for high performance are covered in greater detail in Chap. 9.

5.4.2 Visibility

The very nature of snow removal creates a potential hazard for both the driver of the plow vehicle and other vehicles on the road in the form of the snow cloud produced as dry snow is picked up by the plow and cast to the side and the turbulent wake generated by the plow vehicle as any residual snow is entrained and lifted into the air. One of the most common accidents involving snowplows results from motorists running into the rear of a plow truck because the truck was obscured by the snow cloud. Attempts have been made to reduce the frequency of such accidents by installing deflectors on the rear of the truck to created a downwash to expose the truck markings. Various lighting configurations have been investigated, but little visible light penetrates a snow cloud, and lighting devices are often covered with snow. The plow truck driver's visibility is improved when snow is prevented from blowing over the top of the plow. The potential for blowover increases as plow speed increases, and one solution is to plow at slower speeds. Efficient moldboard designs will enhance the smooth flow of snow along the plow surface without overtopping, and high moldboards will also reduce blowover. Of course, this does not prevent all snow from reaching the windscreen. Heated wiper blades help to prevent snow buildup, and heated windscreens have also been used. A specialized window developed for the marine market, the "clear-sight window," is used in Europe and Scandinavia. This consists of a spinning transparent disk in front of the truck windscreen which flings off any snow or ice by centrifugal force.

5.5 Other Equipment

5.5.1 Graders

As described above, these units are equipped with a center-mounted blade which can apply downpressure for cutting ice. Graders are one of the most common pieces of equipment in many county, town, and rural areas because of their year-round utility. They travel relatively slowly while plowing snow or ice, and therefore are not suitable for heavily traveled or high-speed roads. They are frequently used for clearing shoulders (berms) of snow. Their versatility is compounded when they are equipped with a front-mounted plow and wings for snow removal.

5.5.2 Front-end Loaders

These units, when introduced, were equipped with a bucket on the front lift arms. They are still used in this configuration for loading salt and sand into spreaders, and for moving bulk materials around a yard. However, when equipped with a front plow, they are highly maneuverable snowplows. Articulated loaders, which can swivel their front half about a pivot, have a very short turning radius and are often the unit of choice for clearing snow in cul-de-sacs and other constricted locations.

5.6 Sidewalk Snow Removal

Street snowplows, even pick-up truck plows, are too wide for the usual sidewalk. Small, narrow snowplows are made for this purpose, using either wheels or rubber tracks. Both displacement plows and rotary plows are available.

5.7 Snow Loaders

Rotary plows are commonly used to load snow into trucks for transport to disposal areas. The conventional rotary plow can only cast the snow to a truck alongside, thus restricting traffic during the operation. A recent development in Japan incorporates a chute over the top of the plow carrier to carry the snow to a truck following behind.

Before the development of rotary plows, continuous bucket loaders were used for loading. Though much slower than rotary plows, they are still used. Snow can be carried to trucks following directly behind, as with the rotary unit described above, or the snow can be diverted to the side.

6
Weather and Pavement Condition Intelligence

6.1　It's a War Out There

Fighting Old Man Winter has been likened to fighting a war. In Chap. 2 we used a military analogy and said, "know your enemy." Carrying the analogy further, follow the advice of American Civil War General Nathan Bedford Forrest and "get there firstest with the mostest." To know when and where to send your cavalry with the horsepower behind the snowplows, you need two essential types of information: when your enemy will strike, and what his weapons will be. Getting them is the task of the intelligence service: road weather information systems (RWIS) and their spies deep within enemy territory sending back weather information. The better the intelligence, the more certain the victory, and your troops can add another streamer to their battle flag.

6.2　Sources of Weather/Climate Information

There are many spies ready to provide you with essential and timely intelligence on what your enemy is planning. It's up to you to find them and listen attentively to what they have to report. This chapter is your field manual, a guide to these sources.

6.2.1 Government Sources

Weather has such a major impact on agriculture, construction, travel, infrastructure, and people that probably every country has a national organization established to track and record the daily atmospheric events. The National Weather Service (NWS), currently an agency of the National Oceanic and Atmospheric Administration, is the primary source of weather forecasts and warnings for the entire United States. It generates short-term forecasts and long-range climate outlooks and issues warnings of tornadoes, hurricanes, flash floods, severe storms, and other life-threatening events. It is expanding the network of modern NEXRAD (Next Generation Radar) Doppler radars, which can discriminate between rain and snow cells in the atmosphere over a radius of about 150 miles to provide early notice of advancing snowfall events. The agency also operates a family of high-altitude weather satellites and low-altitude spacecraft for tracking weather patterns.

NOAA Weather Radio currently broadcasts continuous National Weather Service weather information 24 hours a day across the United States from 400 FM transmitters on the following seven frequencies in the VHF band:

162.400 MHz

162.425 MHz

162.450 MHz

162.475 MHz

162.500 MHz

162.525 MHz

162.550 MHz

6.2.2 Private Sources

Though all of the products generated by the National Weather Service are available to highway agencies, the forecast information is often too general to apply to the road microclimate. Detailed, timely forecasts tailored to the needs of highway maintenance departments can be obtained by contracting with private meteorologists or meteorological organizations. These specific forecasts, accompanied by frequent updates when necessary, enable the deployment of crews before precipitation begins. To get the most benefit from an anti-icing strategy, timing is very important. Another advantage of specific forecasts can be the reduction in the number of false call-outs. The savings in materials and labor/equipment time in a single storm may well pay for a meteorological services contract.

6.2.3 Agency Sources

State climatologists (there is an office of that name in every state) can provide historical records for most regions of the country.

6.3 Road Weather Information Systems

Over the last few years, it has become apparent that the road microclimate is extremely variable and differs greatly from that at the sites where weather observations have traditionally been made. Government and private agencies have usually been more concerned with large-scale meteorological events or with information on events affecting specific applications such as agriculture or aviation. This has led to the installation of instruments to monitor several aspects of the road environment and the establishment of communication systems to send this information to a central point for analysis and use in decision making (see Fig. 6-1).

6.3.1 What Is an RWIS?

Simply stated, a road weather information system is a collection of instruments to measure the meteorological and environmental variables that have been found to have the greatest value for microclimatic assessment, and the communication system necessary to collect this information and transmit it to a central location. The observations of the current state of the road and the atmosphere made by the RWIS and by the larger national network are analyzed to make forecasts of the expected weather and road conditions. The accuracy of forecasts specific to the road environment over a limited region is heavily dependent on RWIS observations made in that region.

6.3.2 System Description

An RWIS has three major components: sensors; the electronics for processing the signals generated by the sensors; and the communications facilities that connect the sensors to the processing center, transmit the sensor data to a meteorological forecasting center, receive the forecasts that are made, and finally to communicate the information to the snow removal forces and to the public.

Sensors. The pavement surface condition and the weather in the vicinity of the road must be monitored to determine when conditions are likely to lead to a reduction in friction and to provide the information necessary to choose a counterstrategy.

Figure 6-1. Meteorological tower on roadside for RWIS atmospheric measurements.

Pavement Ice Detectors. Pavement surface condition is monitored by small metallic probes on the surface of an epoxy block which is inserted in the wear course of a pavement so that the sensing elements are flush with the pavement surface. Usually two variables are sensed: the surface temperature and the concentration of a chemical that may be on the surface. Temperature is sensed by thermistors, small solid-state semiconductors that change their resistivity with temperature. Chemical concentration is sensed by measuring the electrical conductivity of the liquid film bridging two small probes. The conductivity of pure water is very low and represents a baseline level. When a chemical compound dissociates, i.e., breaks up into its constituent ions, the electric charges carried by the ions will produce a current flow when a voltage is placed across the probes. The more ions there are, the greater the current flow. This provides a measure of the chemical concentration. Not all chemical compounds dissociate completely into their ionic state. Those that do, such as the so-called ionic salts like sodium chloride, calcium chloride, and magnesium chloride, will have high conductivity that is proportional to their concentration in an aqueous solution. Thus the amount, or concentration, can be determined by measuring the current flow. Others, such as organic compounds like the acetates, formates, and urea, will dissociate incompletely and to varying degrees; thus their conductivities will be lower and, more importantly, the extent of dissociation will vary from compound to compound. It is readily apparent that a chemical concentration sensor cannot be calibrated to provide an accurate value when two or more types of chemicals may be present. However, conductivity does provide a means for detecting whether water or ice is present because of the significant differences in conductivities between the two.

The passive devices just described have several limitations, the chief of which is their inability to report the true freezing point of any surface moisture or to determine with some precision whether the surface is wet, frozen, or dry. They merely respond to the environmental variables and report what they find. Another class of sensor, called active devices, has been developed in the last few years. In these devices, a small region of the sensor is heated, then cooled below the freezing point of water. A temperature probe incorporated in the heated region responds to the temperature variations during the heating-cooling cycle, detecting the point during the cooling phase when the liquid freezes. By this means the effects of any type of ice control chemical can be determined based only on its freezing point, not on whether it is an ionizing substance like salt or a partially ionized compound like CMA.

The temperature of the pavement surface is influenced not only by the air above it but also by the flow of heat below the pavement. Recording the temperature at depth, usually 18 in (46 cm), will give an indication of the direction of heat flow. Prediction models use the two temperatures in their algorithms for greater precision.

Atmospheric Sensors. The microclimate in the vicinity of a road is monitored by an RWIS meteorological station. Variables usually observed are air temperature, dew point, wind speed and direction, and precipitation (presence and type). Some installations also measure solar radiation and visibility. Variations in pavement surface temperature will track the air temperature with some delay, so frequent air temperature measurements will indicate the temperature trend in the pavement and will be a good predictor of the time when freezing conditions might develop. Dew-point temperature provides warning of the onset of conditions that may lead to "flash" icing, the sudden and very transient formation of black ice when the water vapor just above the road surface condenses on the cold road surface and freezes in a thin film. Wind speed and direction are important because of the potential for snow drifting over the road at wind speeds of about 10 mi/h (4.5 m/s) and blowing snow that can obscure visibility at higher wind speeds. Wind speed and direction are conventionally measured on top of a 10-m (32.8-ft) meteorological tower. Because of height restrictions, however, these sensors are sometimes placed at a lower level, or a convenient existing structure such as a sign gantry may be chosen.

Visibility Meters. More precise observation of obscuration caused by falling or blowing snow is provided by optical instruments which determine the visibility. [Visibility is defined as the greatest distance in a given direction at which it is just possible to see and identify with the unaided eye (a) in the daytime, a prominent dark object against the sky at the horizon, and (b) at night, a known, preferably unfocused, moderately intense light source (Geer 1996).] There are two types of instruments: devices which monitor the brightness contrast between an object and its surroundings, and devices that monitor the attenuation of light through a limited volume of air (Ishimoto 1995). The second type, the transmissometer, is more common. Devices that measure precipitation rate and determine whether the precipitation is rain, snow, or ice use the same principle as a visibility meter but also use the scattering properties of precipitation to discriminate among the types.

Signal processing. Sensor outputs are electrical impulses which must be processed and transmitted to a location where their meaning can be interpreted and acted upon. These outputs must be amplified to avoid signal loss and distortion on their passage to the central collection point. This is accomplished by a remote processing unit (RPU) located as closely as practicable to the collection of sensors at a site.

Communications facilities. The sensor outputs from the RPUs are most commonly transmitted to a central location over the public switched telephone network, using telephone dial-up for polling on a schedule. Dedicated leased telephone lines are used for critical sites requiring frequent polling, and microwave telemetry is used for sites remote from a telephone line. The central location is generally a local or regional maintenance office. There the incoming sensor signals are received by a computer (a CPU or central processing unit) and decoded for presenting data either in digital form or frequently in graphical form on a color monitor. The sophisticated computer programs now available can alert the maintenance manager when certain condition thresholds are reached, for example, a critical temperature such as 33°F (2°C), a low chemical factor, or an indication of ice or other precipitation. Facilities also must be provided to communicate with snowplows and spreader trucks on their routes, with supervisory personnel around the region, and with radio, TV, and other media for dissemination of road conditions to the public.

6.4 Radiometers

The molecular motion of all objects produces radiation, which travels through the air in the form of waves. These waves can be intercepted and absorbed by a sensor and their energy determined. The amount of this radiant energy is a measure of the temperature of the object. (See Sec. 3.4.1 for a discussion of transmission of energy by electromagnetic waves.) This provides a noncontact method for measuring the temperature of a pavement surface. Two regions of the electromagnetic spectrum are used for this: the infrared and the microwave.

1. *Infrared devices.* These are the most common noncontact temperature-measuring devices. They are relatively inexpensive, provide rapid response, and when properly used provide an accurate temperature measurement. Two precautions are necessary for accuracy: The instrument must be in thermal equilibrium with its surroundings at the time of measurement, and extraneous radiation must be prevented from entering the sensor window. Portable, hand-held radiometers can be used to measure spot temperatures to provide the information for choosing a treatment strategy, but if they are carried in a warm vehicle, they must be allowed to reach the outdoor temperature before a reading is made so that thermal shock does not lead to an invalid reading. If the instrument "sees" anything other than the pavement, the radiant energy from that source will corrupt the reading. It is necessary to hold the instrument close to the pavement in such a way that it will not cast any shadow. Vehicle-mounted devices, used for obtaining continuous temperature records or for rapidly making spot measurements over a road network, will always be close to the surrounding temperature once the vehicle has left a warm garage, but care must be taken to mount them where only the pavement is in their view.

2. *Microwave systems.* No practicable microwave pavement temperature sensor is manufactured yet, although research has been conducted in a number of countries to explore the use of this region of the electromagnetic spectrum. Mention is included here because there are good scientific grounds for effective utilization of this part of the spectrum for temperature measurement and surface condition assessment.

6.5 Thermography (Thermal Mapping)

Embedded pavement ice detectors make point source measurements. Pavement surface temperatures may vary as much as 13°F (7°C) over a road network because of exposure, elevation differences, soil conditions, subgrade depth, and pavement variability. Since it is impractical to install ice detectors wherever these differences exist, if indeed they are known when making the installations, thermal mapping has been developed to quantify the variation in surface temperatures. Surface temperatures are measured under three different weather conditions using infrared radiometers. This is done rapidly using vehicle-mounted equipment and recording the data on dataloggers for later processing. The processed temperature data are presented graphically on color-coded profile maps which readily show the temperature variations existing over the road segments. Areas that will have a high probability of ice formation can be identified with good accuracy. Information derived from thermal mapping can also be useful in selecting sites for installation of embedded ice detectors.

7
Decision Making for Snow and Ice Control

7.1 Introduction

The quality of decision making is generally in direct relationship to the quality of information that forms the basis for decisions. Information sources range from our senses of hearing and seeing and our experience to sophisticated, computer-based road weather information systems. The *systematic* use of *all* available information is essential for effective decision making.

7.2 Decision Criteria

A vast array of information is available to the maintenance manager. Avoidance of information overload requires selectivity. Listed here are what many consider the important criteria for informed decision making. Each is discussed in some detail in the sections that follow.

- Air temperature
- Pavement temperature
- Character of a snow and ice event
- Time when the event starts

- Event intensity
- Character of the precipitation
- Event duration
- Wind speed/direction
- Time of day/season
- Pavement friction

7.2.1 Air Temperature

For many years air temperature was the sole basis for snow and ice control treatment decisions. Current technology and experience suggests that air temperature, by itself, is not the best factor to use for decision making. Knowledge of pavement temperature allows more effective treatment decisions to be made. By combining air temperature with personal knowledge and other data, estimates of pavement temperature can be made. More precise information is obtained by surface temperature measurement, as described later.

Air temperature is relatively simple and inexpensive to measure. Most maintenance facilities have thermometers mounted outside a window. Proper thermometer selection, siting, and calibration is critical to obtaining meaningful data. Liquid column thermometers, either alcohol or mercury, are more accurate than the dial or bimetallic strip type. Electrical sensors, such as those found in some indoor-outdoor thermometers, are also generally quite accurate. Mounting is important. Thermometers or sensors must be shaded from the sun and protected from other sources of thermal radiation. This may involve using some creative mounting techniques. Many operators and supervisors find it useful to have electronic indoor-outdoor thermometers mounted in their vehicles. The outdoor sensor must be mounted away from heat-emitting engine, cooling, and exhaust system components.

Most measurement devices require calibration. Thermometers are no different. Precision liquid column thermometers that meet specified accuracy requirements are available at scientific supply houses. Use such a thermometer as a reference to periodically check the accuracy of the thermometers that are routinely used. Purchase a precision thermometer about 11.8 in (30 cm) long, with the range of intended use extending nearly full scale. A good range is $-40°$ to $+50°F$ ($-40°$ to $+10°C$). Make sure thermometer users are aware of any differences in calibration.

7.2.2 Pavement Temperature

The working ability of most ice control chemicals is directly related to pavement temperature. As a result, current and predicted pavement temperatures are the *most important* data for deciding on a snow and ice control treatment strategy. A variety of techniques for determining or estimating pavement temperature are identified in Chap. 6 of this manual. In general, the colder the pavement temperature, the more ice control chemical will be required to give equivalent results.

7.2.3 Characteristics of Snow and Ice Events

There is an almost infinite variety of snow and ice event types. These range from icing that occurs in the absence of falling precipitation to heavy, wet snow that can accumulate at the rate of 2 in (5 cm) or more per hour. Dramatic changes in precipitation characteristics often occur within an event. In general, higher accumulations (weight) of snow and ice on the pavement per unit of time require more ice control chemical for equivalent results. Strategies detailed in Chap. 3 can be followed to reduce the amount of chemical used, however.

7.2.4 Start Time of Events

Having good data on event start time is critical if anti-icing strategies are being utilized. Timely response is the key to effective and cost-effective ice control.

7.2.5 Event Intensity

Knowledge of past, current, and future event intensities helps the decision maker to make informed choices on chemical application rates, materials strategy, plowing frequency, resource deployment, and personnel management. In general, less intense events should require less resource expenditure, although this can be affected by other event characteristics (type, ice content, pavement temperature, etc.). Cycle times may be increased (fewer people and less equipment will be required), and chemical application rates may be decreased, consistent with cycle time and pavement temperature.

With extremely high event intensities, other strategies have to be employed. If sustained near-zero visibility is expected or encountered, it may be appropriate for safety reasons to cease operations and allow personnel to rest. In short-term, low-visibility situations, it is probably wise for snow and ice control vehicles to get well off the highway and wait until visibility improves. Chemical application may be of little benefit during periods of intense precipitation. Chemical applications made before or very early in the event will assist in the removal of pack, which is almost unavoidable with sustained high intensity events.

Information on event intensity is becoming much more accurate and timely as the National Weather Service's NEXRAD (Next Generation Radar) radar installations are becoming operational. Also, many private-sector vendors are ready to provide this information to subscribers at modest cost.

7.2.6 Character of Precipitation

Knowledge of the event character is used by the decision maker in making judgments about treatment type and application rate. Event character is usually defined in terms of water or ice content. Terms like "dry" snow and "wet" snow are commonly used. Higher ice content events require higher ice control chemical application rates for equivalent results. See Chap. 3 for further guidance on application rates.

7.2.7 Event Duration

Accurate prediction of how long an event may last will help the decision maker deploy personnel resources and contract resources most effectively. Often, agency personnel resources are insufficient to maintain a sustained high-level effort. In such situations, a less intensive agency response supplemented by contracted resources may be more appropriate for the long haul. Materials application strategies may be compromised if there is insufficient storage capacity or inventory.

7.2.8 Wind Speed/Direction

Having this information will allow for decisions on the direction of snow casting while plowing and on the potential need for poststorm blowover treatment. Judgments about the rate of convectional warming or cooling of the pavement surface can be more accurately made by utilizing wind speed data.

7.2.9 Time of Day/Season

Inferences about pavement temperatures can be made based on the time of day, solar and sky radiation effects, and season. This helps in deciding ice control chemical application rate and timing. Daily pavement temperature cycles are fairly well understood by most people who are involved in snow and ice operations. Absent radiation effects, pavement surface temperature will generally track air temperature with a time delay of a few minutes to several hours. The influence of solar radiation on pavement surface temperature is dramatic. This effect is more pronounced during the time of day and in seasons when the sun rises higher in the sky. The radiation effects of a clear, cold night sky can depress the temperature of pavement surfaces below that of the surrounding air and provide opportunity for "icing" or "frosting" to occur without direct precipitation.

7.3 Pavement Friction

There are two separate and distinct snow and ice control operations. The first job is to make the highway passable. This is typically accomplished by removing accumulations of frozen precipitation with displacement or rotary plows. The second job (or the first job if there isn't much accumulation) is to provide a level of pavement surface friction that will allow vehicles to brake, turn, and accelerate *reasonably* safely. Pavement friction is the most useful indicator (measured or subjective) of the level of snow and ice service.

7.3.1 Measuring Pavement Friction (see also Chap. 4)

A variety of devices suitable for measuring pavement friction in the snow and ice environment are available. They measure rolling friction rather than static friction. The cost of these devices typically ranges from $150,000 to about $2,000 (1996 American dollars). The more expensive devices have a "fifth," or supplemental, wheel that is lowered to contact the pavement. A known resistance or "slip" is added to the wheel, and the relative pavement "drag" is measured. These devices are useful in all traffic environments, as there is no slowing of the measuring vehicle during the friction measurement. They are also capable of measuring continuously over long segments of highway. This may be useful in identifying trouble spots and other locations that require special attention. Less expensive devices (in the $2,000 range) work in conjunction with cars or trucks with antilock brakes. While the vehicle is operating at the test speed, the brakes are forcibly applied and held for a few seconds. The instrumentation calculates friction from the deceleration and the known characteristics of the braking system. These devices may not be suitable for making measurements in traffic or for providing a continuous friction profile.

7.3.2 Estimating Pavement Friction

Surprisingly accurate estimates of pavement friction can be made by visual assessment. This is a skill that will improve with experience.

Terms like "bare," "wet," "icy," "wheel path bare," and "snow-covered" have generally understood meanings and broadly correlate well with measured friction.

7.3.3 Pavement Friction as a Decision-Making Tool

Measured or estimated pavement friction can be used as a snow and ice treatment decision tool. In the absence of data from a road weather information system, friction assessment gives an excellent indication of the presence or absence of ice–pavement bond. This allows treatments to be made only when necessary. When an anti-icing strategy is used, for instance, the chemical, which was applied at low levels, can become diluted so that refreezing begins. Friction measurements are best made frequently when conditions of this nature may arise so that fresh chemical applications can be called for.

7.4 Command and Communication Center

A snow and ice command and communication center (sometimes referred to as a "snow desk") is an integral part of snow and ice control operations at all levels of government. The center may range from a well-staffed multifunctional high-tech operation covering a wide area to one person using a telephone and radio for a town or village. Available resources and the magnitude of snow and ice control operations generally dictate the staffing level and duties of the center. Here are some typical functions of command and communication centers.

7.4.1 Weather and Road Condition Data

Road and weather information is essential to providing a high level of service. People and sensing equipment should be monitoring approaching snow and ice events and present road conditions. Technology provides various types of sophisticated equipment for this purpose, but many good information sources may be less high-tech but equally helpful. Some of these technological aids and simpler but nonetheless valuable aids are:

- Road weather information systems (RWIS)
- Agency road patrols
- Police agency road patrols

- Volunteer spotters
- Private-sector weather/radar service providers
- NOAA weather products (Weather Band Radio, zone and local weather forecasts)
- Weather information products available through computer data service providers (CompuServe, America Online, Prodigy, etc.) and on the Internet
- Radio and TV newscasts
- The Weather Channel

7.4.2 Use of Agency Resources

The command and control center is the logical source of direction for agency resources. Typical activities include:

- Call-in of the necessary number of appropriately skilled personnel
- Assignment of personnel to equipment and routes
- Activity direction—plowing, chemical spreading, abrasives spreading, application rates, etc.
- Demobilization or additional mobilization when required

In the event of emergencies not related to snow and ice, the center can function as an Emergency Management Control center.

7.4.3 Communication with the Public and Road Users

This activity cannot be overemphasized. Communication is the cornerstone of good public relations. This doesn't just happen—it requires a sincere and consistent effort by the agency. Communication of good information, on a timely basis, is one of those rare win-win-win situations: The agency wins by provoking fewer complaints and hence having more freedom to carry out the required operations, the media win by obtaining timely relevant information to give the public, and the public wins by receiving the information needed to make decisions that will help them to avoid hazardous and delaying conditions. The following forums are successfully being utilized to convey relevant information to the media and the traveling public:

- Direct access by media to road and weather information data
- Timely facsimile transmissions to media

- Recorded phone messages for people calling in
- Highway or travel advisory radio transmissions
- Variable message signing along the highway
- Home page or bulletin board on the computer network that is kept updated
- Kiosk stations at rest areas and other locations.

7.5 Command of Snow and Ice Control Operations

The complexity of the command system for snow and ice control operations is generally related to the size and responsibility of the managing agency. A typical chain of command for a large agency may look something like this:

- Agency CEO
- Regional manager
- Regional maintenance manager (responsible for a group of counties or some other large jurisdiction)
- Resident maintenance manager (responsible for one or two counties, or a similar jurisdiction)
- General supervisor
- Crew supervisor
- Individual workers

The chain of command for a small agency will be much simpler:

- Highway superintendent
- Individual workers

There is considerable variety between those extremes. Regardless of the chain of command, the objective is to provide the traveling public with a level of service appropriate for the highway classification. This involves gathering and interpreting road and weather information, deciding on a site-specific strategy, and implementing that strategy. This is not just a one-time action—it must be ongoing, with appropriate, perhaps different, strategies used during and after the storm event. To the extent possible, individual operators should be given the training, information resources, and authority to respond to the continually changing characteristics of these events. Supervisors should be part of the information resource and the primary point of quality and policy assurance.

7.6 Reporting Requirements for Snow and Ice Control

Reporting the actions taken is required before, during, and after each snow and ice event. The detail of this reporting will depend on agency requirements, but it should be sufficient to ensure that the provisions of the agency's snow policy as outlined in Sec. 8.4 are met.

7.6.1 Pre-event Reporting

Preevent reporting provides the decision maker with a variety of information about the pending snow or ice event, equipment resources, personnel resources, and materials resources. While there is no single format for reporting information, essential elements that should be included are

- *Road and weather data.* It is important that timely road and weather data be delivered to the decision maker. In the event that private-sector forecasting is utilized, the contract should define the format and content of the report, transmission methodology, how far in advance of the events the report(s) should begin, and the frequency of reports before, during, and after the event. In the case of in-house data gathering, procedures to ensure the data reach the decision maker promptly must be built in. This may be accomplished by using combinations of electronic alarms and paging, verbal notifications, and requirements for periodically checking the appropriate databases.
- *Equipment, personnel, and materials resources.* The decision maker should know the status of equipment significantly prior to the snow or ice event. This will allow him or her to determine on a timely basis the appropriate level of personnel resources needed and any outside equipment/personnel rental that may be required. Knowledge of the status of readiness of personnel yields information on how much equipment can be utilized and the possible necessity for obtaining additional people from other sources. Information on availability of materials is required in order to make appropriate treatment decisions.

7.6.2 During-the-Event Reporting

The types of information required before the storm are also required during the storm. Operational decisions in response to changing conditions should be based on solid information rather than intuition. During the operational period, information on road conditions should be reported by operators and patrolling supervisors. Reports on road conditions, progress of operations, and projected road conditions should be prepared by the agency and shared with the media and the public as described previously.

7.6.3 Poststorm Reporting

Poststorm reporting is an important management tool that often is not used to its full potential. The most critical documents for this are the operators' trip tickets or shift reports. The following minimum information should be recorded on the trip ticket:

- Operator's name
- Vehicle ID
- Date(s) and duration of shift
- Description of roads treated
- Beginning and end times of each treatment cycle
- Treatment locations and time if not done on a prescribed cycle
- Type of treatment performed on each cycle or run
- Amount and type of materials used on each cycle or run
- Road and traffic conditions observed on each cycle or run

It must be recognized that it is often difficult for operators to record these data. However, these data are *vital*. Supervisory priority and significant training effort must be focused on these reports if lapses occur. In the near future, technology will allow all of these data to be acquired in a paperless environment. Until then, conventional reporting must be used, and it must be treated as a priority.

Nonthreatening poststorm debriefings or critiques should be held after each event. During these sessions, supervisors and operators discuss successes and failures. This provides a vehicle for continuous improvement of snow and ice control operations.

Statistical data on event type and duration, resource consumption, results, and lessons learned should be accumulated and sent up the chain of command. All this information will assist management in judging the impact of resource level, training, public relations efforts, litigation, and other management issues.

8
Managing Snow and Ice Control

8.1 Introduction

One of the most demanding and costly functions that highway agencies in northern areas must perform is snow and ice control. Whether in state, county, city, or town, snow and ice events put a strain on the capacity and safety of the highway system. The responsible agency must have a plan for minimizing the impact of those events. Also, since snow and ice control equipment can negotiate roads that may be impassable to other wheeled vehicles, the agency will probably be involved in responding to other emergency situations. As a result, the agency should have a plan for responding to both internal and external situations.

8.2 Agency Objectives

It is fundamental to the conduct of a snow and ice control organization that a snow plan be prepared. Questions management should ask in its preparation include:

- What snow and ice control services are to be provided?
- What services are basic?
- What will constitute an emergency?
- Should equipment be specialized or multipurpose?
- Should equipment be usable year round or seasonally?

- What resources are available in the private sector?
- When should private-sector resources be used, and to what extent?
- For cities and towns, what other public services need to be continued during a snow and ice control operation, e.g., refuse collection?
- What other municipal resources are available?
- Under what conditions are these other municipal resources to be used?
- How much funding will be available?
- What is the agency's spending authority?
- What does the community consider most important (for example, bare pavement, plowed sidewalks, environmental protection)?
- What are the physical characteristics of the jurisdiction (road and street widths, grades, traffic patterns, bridges and overpasses)?

8.3 Management Control

An agency's management system is the most influential factor controlling the service provided and costs incurred. Because of the nature of snow and ice control, attention to detailed planning and organization take on added significance. Coping with winter storms has certain characteristics that are unlike other activities. Most importantly, it requires centralization of responsibility. Full responsibility for planning, organizing, and directing the snow and ice control effort should rest on a highly placed official who is given the necessary authority. One of the most important initial responsibilities is overseeing the preparation of a realistic snow plan.

8.4 Snow Policy and Operations Plan or Manual

A policy will establish the principles that govern an agency's actions and outline the procedures to be followed. An operations plan or manual will describe more completely how the procedures are to be implemented. A well-crafted policy may be the best insurance an agency can have in responding to any complaints or adverse actions taken against it. A very useful guide to developing a snow and ice control policy has resulted from the experience of William M. Amundson, Sioux City, Iowa (Amundson 1985). An adaptation follows.

1. A written snow and ice control policy will be easier to use in defense of an agency's actions than an insurance policy.
2. If the policy or level of service is to be implemented through an operations manual, the manual should be referred to in the policy and made a part of it by reference.
3. A record system should be established by which the agency can document that it complied with its policy in terms of level of service and the time at which each component of the policy was completed.
4. A written policy should state that it supersedes all previously issued documents and unwritten policies on the subject.
5. If the agency does no snow or ice removal, salting, or sanding, that in itself is a policy and should be formalized by resolution of the agency's governing body.
6. The written policy should state the specific time the policy is in effect and under what circumstances.
7. The agency should stipulate that the maintenance (public works) department, legal department, or both are to receive accident reports and that there is to be coordination among police, maintenance department, and legal department on actions taken to rectify snow and ice problems when they arise.
8. If the policy states the circumstances under which the normal policy is superseded, all the relevant circumstances should be included in the statement.
9. If the normal policy is superseded by a decision-making process, the decision should be made by agreement of as many of the people who are responsible for snow and ice removal functions as practicable—e.g., in the case of a municipality, the mayor/city (town) manager, police chief, public works director, etc.
10. If the agency has followed practices such as not salting new concrete streets for one or two winters or not removing snow piled at intersections or on medians until all snow routes have been completed, these practices should be spelled out in the policy.
11. The policy should be communicated to local newspapers and radio and television stations so that they can inform the public about the level of service the agency is applying.
12. An agency should consider including in its policy specific bridge deck snow and ice control procedures.

13. An agency should consider a policy based on road-clearing priorities rather than tying the policy to specific time periods or specific snow depths that may or may not be meaningful in certain circumstances. These practices are best amplified in an operations manual.
14. If an agency adopts a level of service which is triggered by some natural event, a means of measuring and recording that event should be established. As a suggestion, municipalities can use records published in local newspapers or broadcast by local TV or radio stations to establish or verify the weather conditions. Counties and state agencies can make use of the same resources.
15. An agency should not adopt a policy that is so unrealistic that it can't be complied with except under unusual circumstances.
16. A snow and ice control policy should include a standard complaint procedure, especially if complaints will result in a deviation from the normal policy. As a policy matter, it is desirable to identify who should receive complaints, the circumstances under which complaints call for deviation from the normal policy, and the means for verifying complaints.
17. The wording of any official announcement made to initiate a specific action to facilitate snow and ice removal, such as a restrictive parking ordinance, should be clear and concise, and an official record of its implementation and termination should be kept.

8.5 Some Legalese for Keeping Out of Trouble

Implementing a snow and ice control policy with care and diligence will go a long way toward avoiding legal pitfalls. What follows is not a definitive legal treatise but an explanation of some terms the legal profession will toss around at the drop of a pad of yellow paper. It will not make an attorney out of the most sincere and skilled maintenance operator. Competent legal advice must be obtained for actual situations.

8.5.1 Tort Liability

A tort is a civil wrong or injury. The purpose of a tort action is to seek repayment for damages to property and for injuries to individuals. These factors must exist for a valid tort action:

1. The defendant must owe a legal duty to the plaintiff. (A duty is an obligation, recognized by the law, requiring the actor to conform to a certain standard of conduct for the protection of others against unreasonable risks.)
2. There must be a breach of duty, that is, the defendant must have failed to perform or properly perform that duty.
3. The breach of duty must be a proximate cause of the accident that results. (A proximate cause is one that in a natural and continuous sequence produces the injury, and without which the result would not have occurred.)
4. The plaintiff must have suffered damages as a result.

8.5.2 Negligence

Negligence is the failure to exercise such care as a reasonably prudent and careful person would use under similar circumstances. The essence of negligence is the adequacy of performance. There are two ways in which one can be judged negligent: (1) wrongful performance (misfeasance) or (2) the omission of performance when some act ought to have been performed and was not (nonfeasance).

8.5.3 Standard of Care

If conduct falls below a reasonable standard of care, the responsible persons and/or organizations may be held liable for injuries and damages that resulted from such conduct. When a potentially hazardous condition exists, the reasonableness of action must take into account these factors:

1. Gravity of the harm posed by the condition
2. Likelihood of harm
3. Availability of a method to correct the situation
4. Usefulness of the condition for other purposes
5. Burden of removing the condition

Many sources of information may be introduced in court to aid in establishing the prevailing standard of care. The most important is the agency's own guidelines and policies. Regulations adopted by the agency may define in some detail the minimum requirements that a reasonable person would follow. Sources of information include:

1. Agency directives and policies
2. Directives of a superior agency, e.g., a federal/state or state/local agency
3. Guidelines and policies of other agencies (to demonstrate the state of the art)
4. Guides developed by national and professional organizations such as the American Association of State Highway and Transportation Officials (AASHTO), American Public Works Association (APWA), Institute of Transportation Engineers (ITE), National Association of County Engineers (NACE), American Society for Testing and Materials (ASTM), or Society of Automotive Engineers (SAE)
5. Engineering texts and manuals
6. Professional journals
7. Research publications
8. Opinions of expert witnesses

8.5.4 Notice

The duty to correct a dangerous condition or take other appropriate action arises when notice is received. This is generally considered to occur when a report is filed with an agency of the jurisdiction having responsibility. Once informed, the agency may have an obligation to respond.

Constructive Notice. A duty to act may arise when the agency should have known of the existence of a situation. It may not be necessary to prove that actual notice was given—for example, a recurring melt/freezeback location where a number of accidents have occurred that is not treated (Lewis 1983).

8.6 Other Considerations in Forming a Policy

Many other unique situations facing an agency can contribute to the content of a snow policy. Several are described in the paragraphs that follow, but don't consider this exhaustive—consider it only a starter list.

8.6.1 Snow Ordinances

Snow ordinances should be realistic and logical so that the public can understand and obey them. If they are, they will be effective in providing safety, convenience, and cost saving to both the citizen and to the highway agency. The ordinance should designate snow routes that will require special attention from both the public and the maintenance forces to enable safe and unimpeded travel during a snow and ice emergency. On snow routes, the use of chains or snow tires should be required, abandoning of vehicles should be prohibited, and priority should be given to emergency services and public transportation. Most snow ordinances prohibit street parking during specific times during a snow and ice event. Parked cars pose one of the biggest headaches for snow and ice control personnel, and many ordinances are aimed at reducing this problem. Regulations that prohibit street parking at all times on designated snow routes, permit parking on alternate sides of the street using calendar odd and even dates, prohibit all-night parking, and become effective after a specific amount of snowfall or after a directive by a government official are examples.

8.6.2 Route Prioritization

It's impossible to service every highway, road, or street at once, of course. The most probable danger points in the road network, such as intersections, hills, bridges, school areas, and railroad crossings, should receive attention early in the storm. A high priority should be given to the roads carrying the heaviest traffic in order to avoid the major problems that bottlenecks can cause for snow removal forces and to provide safe roads for the greatest number of motorists. In rural areas, school bus routes rate a high priority. Snow routes or other emergency routes also must receive early attention. If the storm gets ahead of crew stamina or equipment capabilities (including breakdowns), a guiding principle should be that the fewest people are inconvenienced or endangered. In most instances, this means allowing some routes to go temporarily untreated in order to concentrate on the routes carrying the heavier traffic.

8.6.3 Road Closures

The authority to close roads is usually vested in police agencies and high-level government officials. However, in situations of immediate danger, the highway agency should first protect the public and then secure proper authority for the closure. The most common reasons for snow- and ice-related closures are lack of visibility and passability.

Many highways in the snow belt are generally perpendicular to prevailing winds. This makes them candidates for poor visibility, with and without falling snow. As the susceptible roads probably are known, close cooperation between the police and highway agencies is required to implement appropriate closures on a timely basis. Visibility sensors, either independent or associated with road weather information systems, can provide early detection of low-visibility situations. By integrating these sensors with variable message or fixed message signs, the traveling public can be warned of a dangerous situation and directed to a safer course of action.

Impassable roads are usually associated with high snow accumulations on the road, flooding, extremely slippery conditions, accidents or stranded vehicles on the road, and fires and other emergency situations where the road is being used as a platform for emergency response. Detour and contingency plans should be established for all roads. As impassable roads can occur at any time, those plans should be reviewed frequently and be made part of the agency program.

8.6.4 Interagency Cooperation

In order to provide the traveling public with the safest roads possible, the highway agency must have active cooperation from other agencies and groups. This cooperation usually does not happen automatically—it takes significant effort on the part of the local highway manager.

8.6.5 Police Agencies

Cooperation with local police agencies offers much mutual benefit. Police agencies usually do more frequent patrolling than the highway agency. As a result, they can inform the highway agency of potential and active conditions that may affect the highway system. The highway agency is usually called to assist the police with road closures, lane closures, and maintenance and protection of traffic associated with accident and other emergency situations. The highway agency often has communications-capable vehicles on the road. They can alert the police agency of important situations and keep an eye out for vehicles or people that may have been involved in recent major crimes.

This cooperation requires nurturing and *training*. Each agency has to have a clear understanding of its particular responsibilities and have *respect* for the other agency. In most cases, the police agency has the responsibility for reporting potentially hazardous highway conditions to the highway agency. The highway agency usually has the responsibility for determining and executing the appropriate response to the reported condition. In some cases, the police agency will direct the treatment it feels is appropriate. This is not always consistent with good maintenance practice. The police agency should train the highway agency in situation-reporting techniques and procedures to follow if a criminal is encountered. Similarly, the highway agency should

Managing Snow and Ice Control   **141**

make the police agency aware of the various treatment strategies and their application. The forum for communication among the agencies has to be well planned and executed.

A snow ordinance must be backed up by enforcement in order to be effective. This requires police cooperation. The snow and ice ordinance should permit the police to issue fines and to tow or impound abandoned or stuck vehicles. The police should be responsible for coordinating the early release of workers because of snow and ice conditions. Without coordination, the streets can become clogged to the point of gridlock. Even the best-planned snow and ice control operation will come to a standstill then.

8.6.6 Other Agencies

Cooperation among agencies performing similar tasks is vital at the local level. With the trend toward leaner maintenance organizations, there are very few backup people and pieces of equipment available. This will inevitably result in an occasional critical shortage where help from another agency is required. Cooperating agencies should know each other's resources and have a *written* agreement in place that defines the value of materials, equipment, and services that may be provided on an emergency or cooperative basis. To eliminate cash flow, resources may be traded until the breakeven point is realized.

Cooperative partners can be agencies other than highway agencies. Institutions, prisons, school districts, colleges, park agencies, and hospitals all have some snow and ice control resources. Items traded may include equipment, plowing and spreading services, sand, chemical deicers, mowing, paint striping, litter pickup, pothole patching, paving, materials storage, personnel and equipment quarters, computer services, machine shop services, etc. As indicated earlier, it is very important to have written agreements and a written record of all transactions.

If there are no snow and ice partners available, standby equipment rental contracts with the private sector should be in place for the winter season. In most cases this will be significantly more costly than partnering.

8.6.7 Emergency Management Office

A good working relationship with the local Emergency Management Office is essential. There has to be a clear understanding that agency snow and ice equipment generally will not be available for emergency deployment during and immediately after snow and ice events. In severe situations, the Emergency Management Office may be able to acquire resources from more distant locations to assist the local effort.

8.7 Level of Service

The level of snow and ice control service is defined in terms of road passability and pavement friction. This is controlled by the type and frequency of snow and ice control operations. In general, the greater the resource investment per unit of highway area, the higher the level of service. A high level of service is characterized by little snow accumulation before removal, absence of ice–pavement bond during a snow or ice event, and a rapid return to a "wet" or "bare" pavement after the event ends. A low level of service would involve snowplowing once after the event is over. In this situation, the road could be "passable" but "slippery" much of the time.

All agencies have finite resources that generally do not allow the highest level of service on some, maybe all, jurisdictional roads. The agency must then prioritize its efforts. The most common basis for prioritizing response efforts is traffic volume. The idea is that lower levels of service on lower-volume roads will be inconveniencing fewer people. In urban and suburban settings, streets serving bus routes, hospitals, firehouses, schools, and similar important places are often given a high priority.

Motorists tend to adapt to the level of service provided fairly well, as long as it is consistent from event to event and year to year. However, some agencies try to acclimate drivers to winter driving by providing a higher level of service early in the season, then reduce the level of service as the season progresses. Drivers in a more rural environment tend to be less demanding than those in an urban environment. This probably has to do with expectation and the long delays that can occur when high-volume highways exceed their conditional capacity.

The key element in the level of service equation is the cycle time of plow/spreader trucks. Cycle time is the time it takes for the truck to return to where it started while performing its snow and ice control function. Factors that influence cycle time include (1) location of stockpiles relative to the route, (2) traffic volume, (3) posted speed limits, (4) grades, (5) number of stop/backing maneuvers required, (6) highway geometrics, (7) traffic control devices, and other unique situations. A high level of service can generally be provided with a cycle time of 1 to 2 h.

Agency policy on the use of ice control chemicals has a major impact on the level of service that can be achieved. Timely use of ice control chemicals, at the proper application rate for the condition, in conjunction with effective plowing operations will yield a high level of service. The exclusive use of abrasives or abrasives/chemical mixtures will result in a lower level of service (overall pavement friction).

8.8 Snow- and Ice-Related Accidents

On state highway systems in more severe areas of the snow belt, snow and ice accidents (defined as accidents that occur when the road surface contains snow, slush, or ice) represent as much as 26 percent of accidents annually. This is certainly significant, since snow and ice conditions occur only a small proportion of the time. Road crews do their best to keep roads safe, but accidents still happen. Newer vehicles have many features that make them safer and more capable in winter driving situations, including front-wheel drive, all-wheel drive, positive traction differentials, radial tires, effective tire tread designs, and antilock brakes. One factor contributing to accidents may be that drivers expect more from these features than they are capable of providing. Most accidents during snow and ice conditions could be prevented if drivers chose to travel at a speed appropriate for the road conditions and allowed greater spacing between vehicles.

Highway agencies can help reduce snow and ice accidents. One of the best ways to do so is to provide the highest level of service possible given the available resources. To a large extent, this simply involves doing the job smarter. Some modest investment in technology and training may be required to make this happen. Some of the technology options are discussed in Chaps. 3 and 6, and training options are discussed in the next section.

Snow and ice accidents involving snow and ice equipment are only a very small part of the picture. The New York State Department of Transportation (NYSDOT), for example, averages about 25 "incidents" per million miles of highways plowed and treated (NYSDOT vehicles typically travel about 7.5 million miles each year while plowing and applying ice control materials). Most of these incidents involve contact with roadside features like guiderails, bridge rails, curbing, etc. However, one or two fatal accidents occur each year in which vehicles collide with snowplows. These are usually opposite-direction accidents in which the private vehicle has lost control. Although the risk is relatively small, agencies should be sure their vehicles have properly functioning warning lights and other visibility-enhancing features. Operators should be trained in safe plowing techniques and collision avoidance.

Snow clouds generated by truck plows and snowblowers often obscure them from other vehicles. Sometimes it is possible to choose the direction in which to cast the snow so as to avoid an obscuring snow cloud. In situations with moderate snow and strong wind, the snow should be cast downwind if possible. Airfoils on the rear of plow trucks direct some snow downward and help keep the rear of the truck clear. Snowblowers are usually operated after precipitation has stopped. If sight-restricting snow clouds are generated by snowblowers, a safety backup ("shadow") vehicle should be used.

Variable message and condition-specific signs may be employed to warn motorists of hazardous situations and inform them of air and pavement temperatures. In order to remain credible and effective, these signs should be used only as long as the hazardous condition exists. Transmitting information concerning hazardous conditions and air and pavement temperatures over highway or travel advisory radio may alert drivers to be on the lookout for hazards.

8.9 Training

As an emergency service, snow and ice control activities cannot be confined within the normal operational schedule. These activities are triggered by random events of nature. An effective plan actually includes several plans. It begins with an inventory of resources—numbers of personnel, equipment, and materials. Training starts with instructing personnel about winter storm procedures, chains of command, crew call-out procedures, union contract provisions, deployment of personnel, inter- and intradepartmental relationships, safety protocols, and rules and regulations. Employees should be trained as the department's public relations ambassadors to the public, with a helpful, not adversarial, relationship. Multipurpose skills training should be provided for backup and relief situations.

Annual refresher training is necessary at all organizational levels. Initial training should be in group sessions so that individuals get to know the people they will be working with during snow and ice control operations. Sessions should be short at first to avoid boredom or resistance to attending because of other duties. Use 35-mm slides, videotapes, and motion picture films to review previous storms and techniques. Review all written policies and procedures. Training of new participants can include having them view computer-generated programs. These programs are also an effective tool for retraining personnel who have made mistakes in their duties. Do not forget to include backup operators. Experienced operators should be used as trainers, particularly for on-the-job training. Participation in snowplow "roadeos" (competitions that demonstrate crews' skill in performing difficult snow and ice control procedures) provide not only skill building but an opportunity to see how other departments do their work. Attending local, regional, and national seminars on snow and ice control provides an opportunity to discuss and compare techniques and experiences.

8.10 Preseason Preparations

Prepare a plan for dealing with the first snow and ice event. Check material and equipment inventories. Make sure that communications channels are functioning, and that traffic rules and regulations are in place, ready to be invoked on ice- or snow-affected streets. Training and indoctrination of personnel should be complete. Analyze dry runs of various operations and make last-minute adjustments. Make sure that all items of equipment have been serviced and fitted with the necessary attachments, and that multipurpose equipment has been refitted for winter duty. Verify that contracts have been signed with suppliers of materials and equipment.

Generally work crews are scattered over a wide area at the beginning of a winter storm, not concentrated as they would be on a construction site. Consequently, quick control becomes more difficult, and the ability of crews to respond is of added significance. Compounding the problem, not only are crew members widely scattered, but mobility of supervisory personnel is hampered. This is particularly true during the early stages of operations, the most critical period. Further difficulties can arise from the fact that crew members and machines may have to be assembled from several different departments or agencies. Some problems can be avoided by assigning critical early storm duties to personnel who can readily be mustered.

8.11 Off-Season Preparations

Planning for procurement of equipment and materials should commence in the spring. It is best to solicit tenders for chemicals, aggregates, and equipment and parts in early summer in order to be reasonably confident that the materials will be available by the coming winter. Early in the summer, some vendors may be disposing of materials that were not sold during the previous winter, possibly at reduced prices.

It is also advisable to arrange contracts for rented equipment during the summer or early fall. Arranging to hire contractors' equipment is difficult during a storm. Many contractors will already be committed to other agencies. For those who are not, be prepared to pay premium rates. Specific pieces of equipment with specific components and capabilities should be hired to extend the agency's capabilities. A word of caution: Do not agree to pay for no use, but do pay for making the designated equipment available. It is a major advantage to give the contractor radio communications capability. Contracting for rented equipment has many advantages. It eliminates purchase of single-purpose equipment. It provides additional resources when there are peak demands. The key to successful use of rented equipment is to assign the rented equipment to a crew supervisor and work it the same as you would your own.

After the last storm of the season, put the equipment away as though you were expecting another snowstorm. Be ready for the unexpected, know what repairs are needed, know what has to be replaced, and review police traffic incidents related to snow control operations. Most maintenance people will agree that the off-season provides little enough time to plan and implement changes for the next season.

8.12 Use of Weather Information

A plan for obtaining weather information is an important ingredient in any snow and ice control operation. It provides vital information for strategic planning. Sources of weather information are the local TV station, the local cable station (particularly one that provides The Weather Channel), private weather services, and the National Weather Service in the United States. Contract weather services can provide forecasts of precipitation type, temperatures, winds, storm start time, and duration before the storm, with updates during and after the storm. It is advisable to have the forecasts sent at least twice daily during the winter. The first forecast should be scheduled before the workday begins, and the second before the crew workday ends. Special forecasts should be received 24 h before the event. One of the oldest methods of keeping track of approaching storm conditions is the "jungle telegraph" (now a telephone network), whereby a fellow maintenance manager calls you when the storm reaches his or her location. You, in turn, call another official further downwind from you with your information. This method provides timely and current weather information.

8.13 Communication

There are several types of communication to be considered. One is communication of the latest operational plan to the agency's employees. Tell them how you intend to fight the storm, and what resources you plan to use. Teach them proper radio techniques and proper dispatching procedures. Instruct them in the use of the weather information and how to change operations to adjust to the changing weather conditions.

Another is communications with the public. Local radio and television stations should be used to get your message out regarding your activities before and during a storm. This can be very effective. In exchange for broadcasting your information, give the stations newsworthy and up-to-the minute information. Still another communications medium is an answering machine reached by a special telephone number that the public can call to learn of current operations. Have another telephone number that agency crews can call to obtain information regarding their work schedules. Before the season starts, prepare a snow and ice control brochure that outlines your intended operations. Include telephone numbers and other contact information. Mail it as a stuffer with tax bills or some other routine mailing, or as a separate mailing if necessary. It may be possible to have a public service group hand-deliver your brochure to each home and business in a community.

Another type of communication is frequently overlooked: information from employees to supervisors concerning the employees' physical condition. Supervisors should be absolutely aware of this to head off potential problems. Employees should be on the lookout for telltale signs of deteriorating physical condition—both their own and that of their fellow workers. They must be encouraged to communicate this information promptly to their supervisor. Field supervisors should be aware of proper job performance and proper operation of equipment. Abusive activities should be dealt with immediately.

8.14 Records

Information should be recorded during a snow and ice control operation to serve two purposes: to meet immediate needs, and to provide a historical record for analysis. The primary concern in the immediate need category is to give the operations manager up-to-date data on the status of the work so that he or she can resolve critical problems of task completion, personnel assignment, equipment and material needs, and developing situations.

Types of information which might be assembled and perhaps even plotted at regular periods are

- Percentage of streets cleared, by classification
- Total personnel in the field
- Inventory of equipment and operational status
- Inventory of materials
- Number and extent of breakdowns, and future availability of equipment
- Accumulation of overtime
- Snow accumulation
- Planned operations

Today, fast networked computers, geographical information systems (GIS), and geographical positioning systems (GPS) have added new tools to the manager's toolbox. GIS can show what streets have been chemically treated or plowed, and GPS can show the location of vehicles at that moment.

8.15 Operational Strategies

Ice formation on roads and highways is the first matter of concern to most snow-fighting organizations. The first attack is the application of chemicals and abrasives on ice and snow. This can be applications of sodium chloride, calcium chloride, mixtures of the two, or abrasives alone or in combination with chemicals. Effective spreading of these materials at the proper time, in the proper manner, and at proper locations is the critical first step. Improper and imprudent use of these materials can increase costs and degrade the environment.

Better methods and equipment for chemical and abrasive applications are being implemented each year. The science of deicing must balance public demands for bare pavements with protection of the environment. The goal of keeping roads open and traffic moving must be balanced with proper concern for economics and the environment. Use of more ice control materials is not necessarily more effective. The decision must take into account such factors as type of snow (wet or dry), expected temperature conditions at the time of application and following, anticipated variations at the critical freeze–thaw point, methods of application, and types of material. No longer can public works organizations depend on seat-of-the-pants control measures.

Hit-or-miss methods have been replaced by "hit" planning, which depends on careful scheduling, rigid supervision, and conscientious adherence to prescribed procedures and follow-ups by spreading crews.

Every snow-fighting organization has its own preferences concerning the types of plow units to use and where to place them in the field. The alternatives are numerous, both in terms of the type of equipment assigned to each specific job and in terms of the way the units are utilized. Plowing can be single or tandem, and it can be performed by graders, dozers, and multipurpose trucks such as waste collection units and flushers. Types and dimensions of plows and their cutting edges can be varied; plows can be attached in front or under the body; wheel drive and traction arrangements can be varied; equipment obtained from contractors can be of several types and sizes. Choice of equipment is only part of the decision-making responsibility of the maintenance manager.

8.16 Snow Disposal in Cities and Towns

Not all snow that is plowed to the roadside can remain there. Finding places to put all that accumulated snow often taxes the resourcefulness of snow-fighting organizations. Obtaining dumping or disposal areas where snow and ice can be stored until natural melting occurs can be a problem. Some of the problems are haul distances, traffic patterns into and out of disposal areas, the effect of snow melt on stream flows, and the impact of contaminants in the snow and ice on the environment when the accumulations melt and flow into surface waters or seep into the groundwater. Transporting snow is costly and increases in expense as disposal sites extend farther and farther from central areas from which traffic-blocking accumulations must be removed. Mountains of piled-up ice and snow in parks, playgrounds, and other publicly owned sites may remain well into the spring as dirty reminders of winter's travails.

8.17 Performance Evaluation

Every motorist consciously or subconsciously rates the performance of a snow and ice control organization based on his or her driving experience during a storm. These ratings are not objective, but are perceptions of the conditions encountered in a limited exposure. The crews who have worked hard battling the elements to keep traffic moving relatively smoothly under these adverse conditions will quite likely have a very different perception of the effectiveness of the operations. Like it or not, it is the public's perception of the operations that will prevail. A well-prepared organization will devise a rating system for comparison of performance against some norm or standard.

What standards and criteria should be used for a rating system? Some that have been suggested include cost per inch per lane-mile, duration of storm and time for complete opening of all roads, and temperature and wind conditions as a factor affecting operational performance. The cost of personnel, equipment, materials, and supplies will always need to be considered in relation to the number of operational events and their characteristics. This information can be related to other storms with some comparable characteristics and compared from year to year, taking into account the inevitable fiscal and jurisdictional changes. Do not forget to include the costs for administrative personnel, supervisory staff, and office overhead. It is even appropriate to include spring cleanup.

8.18 Postscript

There are several trends in snow removal that should be noted. The speed of plowing is increasing every year. Many communities are adopting a bare pavement policy for all roads. This is often economically impractical, and careful consideration should be given before adopting such a policy. In some communities, the quantity of chemicals and abrasives used on the streets has been drastically reduced because of the realization that the amounts previously used were not needed. Public concern about environmental damage from excessive chemical use and the availability of equipment that enables more precise application of ice control materials have also played a role.

Communities must not be ignorant of the benefits and costs of present snow removal systems. There are no scientifically established rules or step-by-step manual for removing snow and ice that will apply to every community for every type of storm. An agency's snow and ice control procedures are often based on opinion only, but experience is a more solid basis. Even the seemingly best, most basic ideas and concepts must be adapted to local conditions. Experiment with different procedures and application rates until you discover the best method for a particular problem. If you find something that works particularly well, spread the word. Good ideas should be shared. It is said that you are dumb if you do not learn from your mistakes, you are smart if you learn from your mistakes, and you are truly wise if you learn from the mistakes of others.

Snow and ice control is an expensive and demanding operation. It's an investment demanded by the public, but no evidence of your efforts survives the winter. The reward lies in knowing that your success was based on effective planning and the dedicated efforts of many people.

9
Trucks for Snow and Ice Control

9.1 Introduction

This chapter will focus on the development of proper specifications and purchase considerations for the vehicle, sometimes called the prime mover, used to power snow removal equipment. The variety of equipment used and the variety of conditions faced often dictate different requirements for the prime mover. Understanding the demands placed on the prime mover by the various pieces of snow and ice control equipment will help ensure that the complete equipment package delivers optimum productivity during and after the storm.

Because snow and ice control is seldom a constant, year-round, everyday requirement, the prime movers for snow removal equipment are typically powered equipment or vehicles primarily designed for other tasks. This includes motor graders and front-end loaders typically designed for road construction and maintenance, but the most frequently used vehicle is the motor truck, often configured as a "dump truck" designed for general material hauling. A common thread in equipment and vehicles is the link between snow and ice control, a winter activity, and road construction, a rest of the year activity. This connection, no doubt, has its roots in the days when those building the roads were called on to keep their roads passable during the winter. Developing attachments for existing construction equipment was the most economical means to accomplish the task. Today, the snow and ice control task generally falls to those who maintain the roads, rather than those who build them, yet the lineage of the tools used today is unmistakable.

The frequency and severity of snow and ice is often a significant, if not controlling, factor determining the type, size, and design of response equipment chosen. Road surface type, extent of roads to be maintained, and geographic factors also play an important part in the equation. Locations where winters are long and snow is frequent and plentiful often require custom-made specialized equipment, including prime movers designed specifically for the task of snow removal. Examples include large snow blowers used in mountainous regions and other high snowfall areas, and specialized equipment used in airport applications. In general, though, the trucks and equipment used for most of the world's snow and ice control are primarily designed for other purposes. Maximizing prime mover efficiency and effectiveness in snow and ice activities while retaining the required functionality for the primary application is an important part of the selection and purchase processes.

9.2 Prime Movers

The most frequently used prime mover of snow and ice equipment is the motor truck. The reasons include availability, functionality, and adaptability. Since its development at the turn of the twentieth century, the motor truck has come to serve a multitude of cartage applications. Modern-day truck manufacturers generally offer their customers an extensive array of models with vast numbers of optional components and accessories. Customization of trucks for different uses is at the heart of current manufacturing processes. It is this flexibility of production that makes the truck a good choice as a snow and ice equipment prime mover.

Cost-effectiveness, or cost/benefit ratio, is a major consideration in developing prime mover requirements. The ability to custom select the components of the truck requires consideration of the practical need for an available option or feature set. Certain truck manufacturers specialize in custom-built trucks for the most severe service, including severe snow and ice removal. Such a truck may be the perfect vehicle for road maintenance crews in areas receiving 400 or more inches of snow annually. Yet such an investment, perhaps in excess of $200,000, hardly makes sense for an area which may receive 4 inches in any given year. Selection and specification of the truck or other prime mover must be influenced by the frequency and volume of snow and ice.

The variety of trucks used for snow and ice removal worldwide is nearly as broad as the variety of trucks produced. In short, if you can hang a plow on it, you can, in some fashion or another, plow snow with it. In the United States today, plows can be found on everything from small pickup trucks to city refuse haulers. Surprisingly, under the right conditions, many of these applications make sense. Unfortunately, often an application doesn't make good sense or could, with a little planning, be made far more reliable and cost-effective.

As the most widely used snow and ice equipment carrier used today, the medium-duty truck provides a basis for presenting guidelines for selecting or specifying trucks for use in snow and ice control. The following sections offer information relative to truck selection by major component that can typically be controlled by the buyer of a medium-duty truck. Variations that may apply to other classes of trucks, both smaller and larger, are presented as appropriate (truck classification by weight is given in Table 9-1).

It is important to know that in most applications the truck manufacturer supplies only a part of the final tool. Major producers build only the cab and chassis, leaving the installation of bodies, special equipment and accessories, and snow and ice equipment to secondary truck equipment manufacturers and installers. While large-volume customers such as federal or state governments may award contracts to manufacturers for complete vehicles, the final product will involve numerous suppliers and generally multiple assembly points. Smaller-volume buyers will often be forced to purchase the truck chassis alone, then contract for the installation of related bodies and equipment. This approach requires careful consideration of all components and their sources to ensure that the pieces come together properly and the end product meets the users' needs.

The material presented here is not intended as a comprehensive guide to the engineering and specification of all types and designs of snowplow trucks. The regional evolution of requirements around the world precludes complete detailed coverage of every unique feature

Table 9-1. Gross Vehicle Weight Rating (GVWR) Classification

Class	Weight (lb)
1	Under 6000
2	6,000–10,000
3	10,001–14,000
4	14,001–16,000
5	16,001–19,500
6	19,501–26,000
7	26,001–33,000
8	33,001 and above

known to the snow and ice control industry. This discussion does address the key components in terms common to the truck industry in a fashion intended to assist those charged with the task of snow and ice removal in the proper development of needs and the creation of specifications for the purchase of trucks to meet those needs. Truck manufacturers make it a practice to train their dealer or fleet sales personnel in the development and ordering of trucks for various uses. Dealers in snow belt regions, particularly dealers who market to governments, will generally have sales staffs that are versed in the special requirements of snowplow trucks. These individuals can be a valuable resource in development of truck requirements. Published reference materials providing basic information on truck configuration and specifications include the *Motor Truck Engineering Handbook* by James William Finch (Finch 1993) and the *Fleet Manager's Guide to Specification and Procurement* by John E. Dolce (Dolce 1992). Both are published by the Society of Automotive Engineers, Inc.

9.3 Snowplow Truck Component Considerations

For the most part, trucks are designed to carry cargo. A business specializing in the delivery of goods wants to accomplish this with the minimum of cost and the maximum of profit. Such businesses are the truck manufacturers' largest and most demanding customers and, therefore, often have the greatest influence on the design and the components offered for models produced today. The keys to profitable delivery of cargo are maximizing payload and operating at the lowest cost, including purchase price, fuel, and maintenance. The specific goods carried and the delivery locations will influence the makeup of the truck selected, but the goal remains: Move the maximum material at the lowest cost. While these parameters can be applied to the delivery of chemicals to a slick road surface and the movement of snow to the side of the road, there are conflicts between the design of a for-profit cargo truck and that of a truck whose primary, or even secondary, role is snow and ice control. Recognizing where these conflicts exist and how to adjust component requirements to fit a particular snow and ice control operation will ensure the purchase of a productive vehicle or prime mover.

9.3.1 Frames

The heart of a truck is the frame to which all other components are attached (see Fig. 9-1). The frame is the first component of the produc-

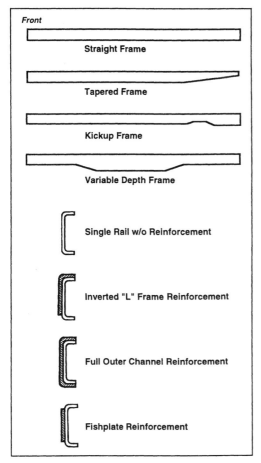

Figure 9-1. Types of frames and their reinforcements.

tion process and the one that is the most difficult to modify after assembly is complete. The first consideration is length. The frame must support all attachments to the vehicle, including the axles, cab, body, and special equipment such as snowplows. The size of the intended body and other equipment and the intended distribution of weight to the axles will dictate the length of the frame and the positioning of the axles, front and rear. Most snowplow trucks are made with two or three axles on a straight frame. The front axle serves as the steering axle, and the rear axle or axles serve as the drive axles and support the majority of the vehicle payload. The positioning of the axles and the distance between axles affect the payload capacity and the ability of the loaded vehicle to comply with weight laws for maximum axle and total vehicle loading of the roadway. Extending the frame length to increase the distance between the axles generally maximizes payload capacity, but the increased length can create maneuverability issues in urban environments.

Most snowplow trucks have body lengths ranging from 8 to 12 ft (2.44 to 3.66 m) when two axles are used or 13 to 16 ft (3.96 to 4.88 m) when three axles (tandem rears) are used. Wheelbases, the distance from the centerline of the front axle to the center of the rear axle(s), range from 130 in to 240 in (3300 to 6100 mm) or more. Maneuverability is frequently an important consideration, in particular when the plowing and spreading involves residential streets. Payload capacity for chemicals and abrasives can be of greater importance for highway departments with long distances to cover with a single truck.

A critical element of frame specification for a snow and ice truck is the fabrication and strength of the frame (see Fig. 9-2). This is where the differences between hauling goods for profit and plowing snow first become apparent. A truck frame must support all loads, yet flex as

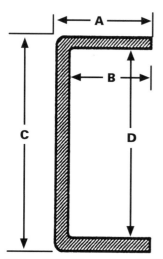

Figure 9-2. Dimension affecting strength and resisting bending moment.

Formulas for flange thickness, section modulus, and resistance bending moment (RBM) for a full C-channel frame; refer to the drawing for the locations of the cross-section dimensions:

Flange thickness, $t = A - B = (C - D)/2$

Section modulus, $SM = (AC^3 - BD^3)/6C$

Resistance bending moment, $RBM = \text{yield strength} \times SM$

the truck traverses uneven terrain. The most common frames are of steel construction. High-strength steels have helped to provide increased capacity with less frame weight. Still, the more steel, the more added weight (lost payload capacity) and cost. To control cost and weight, manufacturers offer a variety of frame choices and frequently design frames specifically to reduce weight in the chassis (see Table 9-2). A common practice is to place more steel, in the form of a reinforcement channel, only in the critical bending area—behind the cab and ahead of the rear axle. This practice is fine if the only loads are from the payload in the box or body on the rear of the truck. However, a snowplow truck will incur additional loads from front plows, and also from side or wing plows if the truck is equipped with them. Sometimes frames are bent or formed to fit other design considerations, such as accepting a larger engine in heavy-duty applications or lowering overall vehicle height in the case of pickup trucks and their light-duty derivatives. These frames seldom are designed to account for the shock and unique loading characteristics associated with snowplowing.

Frames are most frequently rated by resistance to bending moment (RBM). This rating not only covers the bending rating of the frame, but is a good measure of the frame's overall capacity. The higher the RBM, the stronger and often heavier the frame. RBM takes into consideration both the section of the frame and the strength rating of the steel. In years past, lower- or softer-grade steels were used which required a high-volume cross section to attain a high RBM. Today most medium- and heavy-duty trucks use high-strength steel, which requires less cross section to attain the required RBM. RBM should always be the primary rating considered in specifying a truck frame for snow and ice control service.

In nearly all snowplow applications, specifying heavier-than-standard frame pays operational dividends. The ideal frame has a constant cross section and RBM from front to rear and, when the manufacturer offers a choice, should be the highest-rated frame of this type for the

Table 9-2. Typical Frame Specification

Siderail section (in.)	Section Modulus	Yield Strength (lb/in^2)	RBM (in-lb)
10.00×3.06×0.25	10.76	110,000	1,183,600
10.12×3.13×0.312	13.3	110,000	1,463,000
10.25×3.19×0.373	15.9	110,000	1,749,000
14.12×3.13×0.312	21.6	110,000	2,376,000
14.25×3.19×0.373	26.0	110,000	2,860,000

class of vehicle under consideration. Reinforced frames may be considered to further increase RBM, but the reinforcement should be continuous from front to rear to provide equal capacity to support front plow attachments and body loading. In typical truck analysis, longer wheelbases dictate higher RBMs. Specifying the frame recommended for a long-wheelbase truck in a short-wheelbase snowplow chassis of the same model usually results in a suitable product.

A final note on frame requirements is determining the need for a front frame extension for the installation of front crankshaft hydraulic pumps and/or plow equipment. Where snowplow equipment of any significant size is anticipated, extension of the frame beyond the front of the vehicle is best accomplished by extending the total frame, not by add-on extensions. Should add-on extensions be the only available option, as may be the case with some lighter-duty trucks, care should be taken to ensure that both the steel and the fasteners provide strength equal to that of the main truck frame.

9.3.2 Axles

As with frames, truck manufacturers offer a wide variety of options for front and rear axles. And, again as with frames, selecting the right axles for snowplow applications, not standard truck applications, pays dividends. This is true particularly for front axles. In a typical application, a front axle supports a lower percentage of the total gross weight. It is common for the front axle to carry only half the weight of the rear axle in two-axle configurations and perhaps as little as 20 percent in three-axle setups. Introduction of front-mounted plowing equipment quickly changes the load on the front axle. Knowing the extent of this load, which can be anywhere from 1000 lb (454 kg) for a small front plow to 6000 lb (2722 kg) for a large front plow with left and right wing plows, is critical to the reliability and safe operation of the plow truck.

Truck components—tires, wheels, brakes, springs, etc.—are sized to match the gross axle weight rating (GAWR) desired for the front and rear axles and the subsequent gross vehicle weight rating (GVWR) for the entire truck. While it is sometimes possible to mix heavier and lighter components to meet unique needs (the rating is then based on the lightest-rated component), for most applications a balance of all these components to the rating required by the loads applied delivers the best results.

Introduction of almost any form of front-mounted snow removal equipment necessitates an increase in the front GAWR above that required for any other application. Axle loads for a two-axle dump truck with 4-yd- (3 m^3)-capacity body are typically 7500 lb (3400 kg) on the front and 18,000 lb (8165 kg) on the rear. Add a snowplow to this truck and the front axle loads will probably reach 9000 lb (4082 kg) or more. Heavier GAWRs mean heavier components and increased costs. Knowing the intended plow equipment and its weight before specifying axles and GAWRs is critical to achieving a cost-effective plow truck.

Sizing of rear axles typically follows traditional payload sizing criteria. What and how much is to be hauled in all planned uses of the vehicle, along with legal weight limits for the service area, should dictate the sizing of rear axles. In the case of special severe-duty snow and ice trucks, additional axle capacity may be required not because of payload, but to resist the high torque loads associated with pushing large volumes of snow with front and side wing plows.

The secondary nature of snow and ice control frequently leads to the use of dump trucks as the primary carrier, generally because of their versatility for other roadway maintenance functions. Proper sizing of axles and related components to ensure sufficient front GAWR for front plows results in a good multipurpose truck.

4×4 and 6×6 Axle Configurations. Addition of a mechanically driven front axle in conjunction with the standard rear drive is a common practice in trucks used in "off-highway" applications and severe-duty operations such as custom heavy-duty snowplow trucks. Providing power to the front axle distributes tractive effort to more wheels and reduces the potential for wheel spinout in slick or soft terrain conditions. Four-wheel drive (4×4) in two-axle trucks and six-wheel drive (6×6) in three-axle trucks should be considered for certain snow and ice truck applications.

The cost/benefit ratio of adding a driving front axle is frequently high, and therefore this feature is not included in the majority of general duty snow and ice trucks. Adding the feature significantly increases purchase cost and maintenance costs and decreases drivability in normal highway operations. Benefits derived from driving front axles depend on the severity of snowplowing conditions and proper configuration of the feature to the rest of the truck and plow equipment.

Integration of a driving front axle into the truck chassis is typically accomplished with a "transfer case" installed behind the transmission of the truck. The case allows for the division and distribution of power to the front and the rear of the vehicle. A recent variant of this approach eliminates the independent transfer box and incorporates the division of power within the rear axle housing. Some cases are of a two-speed design, allowing for further gear reduction beyond the transmission. This feature provides a range of very low gears for maximum torque multiplication to the drive wheels and should always be considered in heavy-duty plow trucks intended to push large volumes of snow.

Application of different transfer case designs creates vehicles with different operating characteristics. Most transfer cases are designed to provide part-time power to the front axle. This approach gives the driver the ability to operate the vehicle in a two-wheel-drive (all power to rear axle) mode for most applications and engage the front axle only as conditions warrant. Because most part-time transfer cases do not include differential gearing, they should not be used on dry roadways. For proper use of a part-time 4×4 or 6×6 vehicle, the driver must understand when and where to engage the front axle. More sophisticated systems provide full-time all-wheel drive. With these systems, the transfer case includes an interaxle differential and the vehicle is operated with the front axle driving at all times. A more expensive option, all-wheel drive provides distributed front and rear power under all conditions and removes the choice from the driver. Potential use of the vehicle by drivers with little training or understanding of 4×4 characteristics often justifies the added cost of all-wheel drive.

In the worst of surface conditions, the best vehicle will have power to all wheels. However, accomplishing such a power distribution is not a simple task. To operate on dry surfaces where traction is good, a vehicle completing a turn must be able to adjust for or differentiate the different wheel speeds. The typical differential design makes this adjustment by directing power to the wheel with the least resistance. This works well when all wheels receiving power are on surfaces of equivalent friction. Unfortunately, snow and ice conditions often produce unbalanced surface friction, with conditions ranging from spots of dry pavement to water over ice. Thus, even a 4×4 truck with standard or "open" differentials has the potential to spin out as power is directed to the wheel of least resistance—the wheel on the ice. This problem is generally addressed with the installation of limited-slip or "no-spin" differentials. These optional designs can be automatic or driver-controlled and should be considered for at least the rear axle and full-time transfer case when developing a 4×4 snowplow truck. Placing a limited-slip-type differential in the front driving axle is possible, but this is not generally recommended by most manufacturers if the vehicle is to be operated on hard, dry surfaces. It should be given consideration in developing requirements for dedicated heavy-duty snow removal vehicles.

The same loading considerations that apply to nondriving front axles also apply to mechanical front axles. The additional loads created by the plow equipment must be considered in establishing the required GAWR. Often the more economical part-time systems developed for other users such as utility companies and oilfield work do not offer the capacity or features needed to be productive in a snow and ice environment. If a 4×4 or 6×6 truck is needed for snow and ice control, a large investment in a custom truck designed and engineered specifically for the task will typically be required.

9.3.3 Wheels and Tires

Two types of wheels are commonly used on large trucks. One type, commonly called "cast spoke," uses a cast iron center with stamped steel rims mounted at the ends of the hub spokes. Most of the cast hubs are five-spoke, although heavier GAWR ratings will require a six-spoke wheel. The second type in widespread use is the "disc" wheel, which uses stamped steel in place of the cast iron radial spokes. Both designs have their proponents. Cast wheels, the older of the designs, remain popular in the eastern regions of the United States. Disc wheels dominate the market for trucks sold in western areas. Both are used effectively in their respective stronghold territories for snow and ice applications. Only when very high GAWR ratings are required is choice eliminated, as the option of cast spoke is no longer available.

Choosing tires is an important consideration in developing a truck for snow and ice control. All major tire suppliers offer drive-axle tires with aggressive tread patterns for use in off-highway and snow applications. The degree of aggressiveness will affect dry highway performance and tread wear. A good multiple-use tread should be considered when the vehicle is expected to serve multiple tasks. Steering-axle tires for snow and ice use also warrant special attention. Again, tire manufacturers offer a wide array of choices, most of which will work reasonably well for plowing operations. However, when high GAWRs are required [16,000 lb (7257 kg) and above], tire choices diminish. Often the first option is the wide profile or "super single" tire design. While this tire is good for distributing the weight over a wider surface area, it is not always a good choice for snow and ice. Higher-rated traditional-profile tires are available, although at a higher cost, and should be considered based on GAWR and expected road surfaces during typical operations.

In developing wheel and tire requirements for trucks used frequently for snow and ice control, installation of tire chains must also be considered. Clearances between wheel frames and fenders should be checked if chains are to be used. While most trucks with common sizes of tires will accommodate chains, certain tire sizes and profiles may create problems. If chain use is expected even occasionally, this must be considered during the specification of the vehicle.

9.3.4 Engines

Nowhere is the truck manufacturer's array of options greater than in the engines offered across their product lines. The choice of individual engine manufacturer, engine model or series, and horsepower are all available in varying degrees throughout the industry. The diesel engine remains the most common type used in medium- and heavy-duty trucks. Light-duty trucks often offer a choice of gasoline or diesel. In addition, continued interest in, and for some government fleets mandated migration to, alternative fuels has led to the introduction of special fuel engines, including designs that run on compressed natural gas, LP gas, methanol, and ethanol.

Sizing truck engines and the drive trains behind them involves taking into consideration the desired speeds, loads, gradability, and startability factors. Obviously, the power requirements for moving an 80,000-lb (36,300 kg) freight truck over the Rocky Mountains at interstate highway speeds differ greatly from those for moving 20,000 lb (9000 kg) of freight around on city streets. Established formulas, now frequently presented in computer programs, enable the truck salesperson to fit a truck with the right engine and components based on inputs supplied by the customer. Unfortunately, none of these formulas effectively include snowplowing. Still, with some allowances, they are a useful tool in sizing the truck components, including the engine, for the task.

The most important elements in establishing needed engine horsepower are gross weight, frontal area, and expected top speed. The fourth factor in a snowplow vehicle, which is generally unknown, is the load of the snow on the plow. In most applications speed significantly affects horsepower demand. Since most plowing occurs at lower speeds, a reasonable approach is to consider the horsepower allocated to speed, say a top speed with full gross load of 65 mi/h (105 km/h), as available horsepower to plow at lower speeds of 15 to 25 mi/h (24 to 40 km/h). Certainly the severity of the snow conditions typically encountered, introduction of side wing plows, and grades in mountainous regions may suggest a higher horsepower requirement. Cost/benefit ratios should be considered, in particular when considering a move to a significantly more expensive engine class in order to gain a small increase in horsepower.

Engine torque rating is also an important consideration. Torque rating, multiplied through the driveline, establishes startability and gradability of the vehicle. Modern diesel engines often offer high torque ratings, and thus require fewer transmission gears to move the loaded vehicle to desired road speed. A good torque rating is a positive in snow removal whenever heavy snow loading is expected.

Certain engine accessories for snow and ice control vehicles make good sense. In cold regions, electric block heaters and fuel tank warmers help ensure starts in subzero weather. Likewise, heat-transfer devices in the fuel line to maintain fuel temperature above the point where wax crystals form are critical to cold-weather operation. Blowing snow can be a serious problem if it is drawn into the air intake, as it often clogs air filters and chokes engine performance to the point of shutdown. Factory options or aftermarket devices to either draw engine air from inside the engine compartment or prescreen or filter snow out of the air stream before it reaches the primary paper filter are a must in severe blowing conditions. Increased battery capacity and other cold climate options should be considered as operating conditions warrant.

9.3.5 Transmissions and Driveline

The second longest list of options on most trucks involves transmissions and rear axles. Engineered to efficiently move the engine horsepower and torque from the engine to the drive wheels, the transmission, driveline, and drive axles contain the gearing to start the load moving and to carry it to and maintain the desired road speed. Modern diesel engines with higher torque ratings sustained over broader RPM ranges require fewer gears and gear changes to accelerate loaded trucks to highway speeds. However, when developing a truck for snow and ice operations, it is important to recognize that the vehicle will be expected to perform at slower plowing speeds as well as at highway speed when used for other hauling activities. Ensuring the availability of good "plowing gears" to handle heavy snow loads at speeds of 10 to 25 mi/h (16 to 40 km/h) is critical to an effective and responsive plow truck.

Most transmissions used in medium- and heavy-duty trucks are manual shift. Though much improved over the years to ease operation for the driver, these transmissions still require the operator to clutch and manually select the gears based on engine speed and load. The number of gears typically ranges from five to ten, and in some configurations a two-speed rear axle doubles the options provided by the transmission. As previously discussed, 4×4 and 6×6 vehicles often include two-speed transfer cases, again doubling the speeds available from the transmission. Many newer models offer synchromesh gearing, which eliminates the need to double clutch to exactly match engine and transmission speeds. This feature combined with fewer gears can improve drivability and should be a consideration when relatively inexperienced drivers are expected to operate the vehicle in snow and ice conditions.

Automatic transmissions are a popular but costly option frequently used in snow and ice control trucks. Although they do not totally remove the driver from the gear-selection process, they do allow the driver to focus on other driving issues—often an important consideration in snow and ice control operations in poor weather conditions. Modern automatic transmissions offer electronic controls which can be tailored to snow and ice operations, including adjusting shift points to match the loads created by the loaded plow and holding gears for maximum engine performance at slower plowing speeds. Typically, automatic transmissions do not require two-speed rear axles, but can be matched with two-speed transfer cases in 4×4 vehicles to provide a full range of low-speed gears for heavy plowing conditions. Calculation of the cost/benefit ratio for an automatic transmission must include a factor for reduced maintenance cost and the skill level of the drivers. Highly skilled drivers can make manual transmissions the clear cost winner. Yet, for many fleets, the advantages offered by automatics produce a good return on the higher purchase cost.

Rear-axle ratio options are generally matched to the engine operating speed and the transmission's low gear ratio to provide good startability and a fuel-efficient top cruising speed. In developing a truck for snow and ice control, it is equally important to check the selected ratio for operating speeds in the preferred plowing gears. If the vehicle is used extensively for plowing, the engine/gearing match at plowing speed is likely to be more important to performance and driver acceptance than the match for highway-speed travel.

9.3.6 Brakes

The majority of heavy-duty trucks are equipped with air brakes. Medium-duty trucks often are built with hydraulic brake systems, but offer the option of air brakes. The advantages of air-brake systems are apparent from their nearly exclusive use in heavy trucks. In snow and ice applications, optional accessories may be required to keep air-brake systems fully effective and problem-free. Controlling moisture and removing it from the compressed air system is important to the maintenance of the system. This is more critical in the cold conditions of snow and ice removal. Automatic tank drains and aftercooler or desiccant-type "air dryers" are important options for any air-brake-equipped plow truck. While hydraulic-brake trucks eliminate the cold-weather issues of compressed air, they are typically available only on trucks under 30,000 lb (13,600 kg) gross weight, typically the lower limit of size for most plowing fleets in cold climates.

9.3.7 Cabs

The conventional truck configuration in the United States places the operator's cab behind the engine, which in turn is covered by a housing incorporating fenders over the front wheels. The other alternative, the most common in most of the rest of the world, is the cab-over-engine (COE) design, in which the operator's cab sits directly over the engine and the front axle. In typical truck applications, transport of materials, each has its advantages. COE trucks offer shorter overall length for maximum cargo capacity where length is limited by law and increased maneuverability in tight urban streets. Conventional designs are generally less expensive to build, offer more cab room, and have better ride characteristics. Although frequently equipped with snow removal equipment in Europe, COE trucks are rarely used for snow and ice control in the United States or Canada. When they are used, the snow removal application is clearly secondary to the primary purpose, as in the use of refuse trucks for plowing in some major cities. The difficulty of installing plow equipment while retaining easy tilting of the cab to service and maintain the engine is the most common reason for avoiding COE trucks for plowing operations.

Operating a snow and ice control vehicle in what can be the worst of driving conditions is no easy task. Beyond simply guiding the vehicle under difficult road conditions and with poor visibility, the driver must also operate the plows and chemical or abrasive applicators. Providing a suitable work environment helps keep operators effective and promotes the safety of the operator and the public encountered on the roadways.

Once very basic in features and amenities, today's trucks offer numerous ergonomic features. Still, as with other components, manufacturers offer many choices and options to satisfy customer demand. The basics are standard. Selecting the necessary options is up to those developing the truck for its intended application.

When trucks are intended for extensive snow and ice work, consideration should be given to features such as heavy-duty wiper systems with arctic or heated blades, high-capacity heater/defroster systems, high-grade suspension-style operator seat, power steering, a full complement of instrumentation including an air filter restriction indicator and transmission temperature gauge, heated rearview mirrors, and adequate storage for extra clothing and survival gear. While most of the manufacturers that specialize in custom heavy-duty snowplow trucks develop their cabs to incorporate the most effective and necessary features, volume producers of conventional medium- and heavy-duty highway trucks often do not have all the desired items available. When working with these vehicles as primary snow and ice vehicles, it may be necessary to seek aftermarket/add-on accessories to fill certain needs.

9.3.8 Lighting Systems

The installation of almost any front plow will necessitate the addition of auxiliary headlights and front signal lights. And, since the majority of public-use snow and ice vehicles are considered emergency service, some form of warning light package will also be required.

Auxiliary lighting must be installed to provide the optimum night vision ahead of the vehicle for the operator and still provide a light pattern which considers the visibility and safety of approaching traffic. Equipment operated on public roadways can be subject to legal restrictions on the number and placement of lights. While in some cases there are exemptions for snow removal equipment, the specific rules should be investigated and understood.

Veteran drivers' opinions as to the best placement of auxiliary lighting vary extensively. Limited research suggests that the best locations are ahead of and offset from the driver's normal line of sight. Understanding the user environment and gathering input from the drivers before selecting and installing auxiliary lighting is highly recommended.

Warning light systems also vary widely in design and installation. Local and state laws on lens colors and users' opinions as to the effectiveness of different light sources and their position on the vehicle results in extensive choices from suppliers. The most common designs use rotating incandescent beam lamps, often in connection with mirrors, inside a colored translucent housing to create a flashing effect. The second common type of lamp, which is gaining in popularity, is the flashing strobe design, again using a colored lens as required.

The objective of warning lights is to warn others of the location of the snow and ice control vehicle. Thus, they should be placed so as to have the best effect in alerting others. This will vary with the exact application and the conditions. Needs on urban streets will differ from those on rural interstate highways. User input and past practice are two important elements in selecting and installing warning light systems.

Additional or auxiliary marker and signal lights are frequently added to snow and ice vehicles. This is most common on the rear of the vehicle, where the manufacturer's standard frame-mounted stop/turn/tail lights are often quickly covered and obscured by snow. Adding additional sets of stop/turn/tail assemblies higher on the body often improves their visibility for approaching motorists.

Snow and ice vehicles exposed to extensive duty involving corrosive chemicals such as salt and calcium chloride may experience maintenance problems with the lights and electric wiring. Though today's standard lighting harnesses are much improved, some users may consider using custom high-performance sealed lamps and harnesses to improve reliability and reduce downtime due to electrical failures.

9.3.9 Hydraulic Systems

Modern snowplows and ice control equipment rely extensively on hydraulic power to move equipment into position and to drive conveyors and distribution systems. State-of-the-art hydraulic systems are fundamental parts of today's construction and maintenance machines such as motor graders and front-end loaders. The efficient and powerful hydraulic systems that are necessary for their primary tasks make these machines easily adaptable to snow removal equipment. In contrast, the production motor truck has no such ready source of hydraulic power. Aftermarket systems must be added.

The extent of the vehicle's use for snow and ice control duty and the types of equipment to be operated should dictate the degree of sophistication of the hydraulic system. With complexity comes cost. Still, for heavy-duty multifunction vehicles, modern high-capacity systems contribute to the effectiveness of the operation. Systems can be engine- or transmission-driven, but in all but the most limited-use cases should be "live" or available at all times the engine is running. The system should have the pump and oil reservoir capacity to handle all functions simultaneously. This is of particular importance when hydraulic motor–driven material spreaders are used. System capacity should be adequate to allow the plows or dump box to be raised without slowing or stopping the required flow of oil to the spreader drive.

Well-designed systems can be largely trouble-free. However, in severe conditions, hose failures may occur. Any system should include a warning device for the operator, permitting system shutdown before oil loss destroys expensive components.

9.3.10 Bodies

Most frequently, snow and ice trucks are equipped with a dump body. This body type provides the multipurpose, year-round versatility typically sought by those securing snow and ice removal equipment. The body may be a part of the snow and ice control package, used to carry and dispense chemicals or abrasives through spreader attachments, or it may only serve to hold a V-type spreader box installed for the snow and ice season. In developing requirements for the dump box, the nature of all material to be hauled should be considered along with life expectancy. Bodies can be purchased with different thicknesses of sheet steel and designs to match the size and abrasiveness of the materials hauled. If the body is to be used to carry or will be exposed to snow and ice control chemicals, consideration should be given to probable corrosion over the life of the truck. Extra steel thickness and frame designs which ease cleaning and minimize trapping of chemicals will prolong body life. Alternative materials such as stainless steel and aluminum may be considered, but will substantially increase cost.

9.4 Installation of Snow and Ice Equipment

As previously stated, the majority of truck manufacturers build only the basic power platform, called the cab and chassis. While they will develop the cab and chassis to meet many of the needs of the final intended configuration, the final assembly is left to secondary truck equipment suppliers. These secondary suppliers typically specialize in certain markets, building and installing truck equipment to complete the vehicle for a given use. Some larger suppliers focus on installing equipment on a wide range of trucks for a variety of uses. For some large fleets, the final build-out is completed "in house," with the installation of required equipment completed in their own facilities. Only by working with the specialized snow and ice truck manufacturers is it possible to receive a complete, field-ready truck from the point of initial manufacture. Even then, the body and plow equipment is manufactured and provided by other sources.

The integration process introduces an additional array of options in equipment and componentry. Regardless of source or point of installation, the following areas must be considered and addressed in vehicle development.

9.4.1 Hitches

It is the snowplow hitch that allows the plow to be easily attached to the truck. As snowplowing equipment has evolved, the hitch has also experienced changes in design and features. Today most hitches for front plows are of the front frame mount design. This involves plates and bracing bolted and/or welded to the truck frame, with mounting points for quickly and easily attaching the plow. The hitch will typically incorporate the hydraulic lift mechanism for raising the plow once it is attached.

The hitch must be of sufficient design and strength to transfer the loads created by the plow through to the truck frame. This includes the loads of plowing and the loads of carrying the plow when it is raised in transit. Plow size and truck size dictate the design of the hitch. Plow manufacturers will offer a variety of hitches matched to their plow types and required truck sizes.

9.4.2 Wing Mounts

Wing or side-mounted plows require additional attaching hardware. The nose or leading-end attachment of a wing is frequently integrated into the front plow hitch. Elevating or benching wings utilize slide posts at the front and rear. The front slide posts are nearly always incorporated into the front hitch assembly. Again, plow manufacturers develop hitch designs for the majority of trucks used for heavy-duty snow and ice applications. Knowing the hitch manufacturer's desired mounting points prior to truck chassis production may enable the truck manufacturer to build the truck leaving certain areas of the frame clear for the installations that are to be made.

9.4.3 Controlling Costs in Specifying Attachments

Installation of snowplows and spreading equipment is seldom a simple task. Keep in mind that most trucks are not specifically designed and built to accept snow removal equipment. Meanwhile, the variety of truck shapes, sizes, and styles demands that the plow equipment supplier be able to adapt its products in innumerable ways. Specifying special options from the truck manufacturer such as frame extension, engine access hatches, or fixed grille assemblies and clear frame rails behind the cab will help simplify the task of fitting equipment to the truck. These considerations become increasingly important when the truck is expected to carry side wing, underbody, or mid-mount plows. Extensive modification or relocation of factory-installed chassis components by the aftermarket installer can be expensive.

Good installations will take into consideration the vibration and shock loads associated with snowplow operation and the inevitable damage to components from routine use and occasional misuse. While plow trip systems provide a reasonable measure of protection to the plow frames and vehicle, installations which include additional shear points or easily repaired collapsing members in the auxiliary framework will help protect expensive-to-repair truck frames and engine and transmission mounts. Designing an "indestructible" plow vehicle is an extremely expensive, if not impossible, task.

Understanding that developing a snow and ice control vehicle requires contributions from numerous manufacturers, suppliers, and installers is important in receiving a final vehicle that will meet the needs. Snow and ice equipment manufacturers and truck equipment installers who frequently deal with this equipment are the best source of information for developing the final configuration. However, to be effective, they must know the truck chassis with which they have to work. Up-front coordination is critical to a successful marriage of plows to truck.

9.5 Rotary Plow Carriers

Up to this point, this material has focused on trucks to be used with traditional displacement-type snowplows. Rotary plows, or blowers, as they are commonly called, are also a common tool in heavy snow regions. Once routinely mounted on production-line truck chassis, today's rotary designs will typically require custom-designed and custom-built carriers. Many rotary plow manufacturers actually build their own integrated carriers. For those situations where a rotary unit will be fitted, or refitted, to a standard truck chassis, the following points should be taken into consideration.

Rotary plows operate at much lower speeds than do standard trucks. Transmission, multispeed 4×4 transfer case, and axle ratios must be selected to provide an optimum number of low-speed [0 to 5 mi/h (8 km/h)] gear selections. Rotary plow heads, the part that cuts and blows the snow, are much heavier than standard front displacement plows. Front gross axle weight ratings that are high enough to carry the blower head in transport position are a must. A 4×4 design placing tractive effort directly behind the blower head is also a must. Since the rotary plow has its own power source, the carrier engine need not have a tremendously high horsepower rating. The carrier engine should be matched both to the gross weight of the total package and to the drive train to deliver full rated horsepower to the wheels in each of the operating low gears. Transfer case and axle differential locks are vitally important features in an effective rotary carrier.

Requiring the rotary plow manufacturer to provide the carrier is always the best approach. When installing a new carrier under an older rotary, getting the manufacturer's current carrier requirements and recommendations is a good place to start. Since the service life of rotaries often approaches 25 years, the extra time spent on researching the details contributes long-term benefits.

9.6 Reliability and Maintainability

Snow and ice control vehicles, regardless of any of the other tasks for which they are used, are expected to perform from the start of a snowstorm until the roadways are cleared. In the development of vehicle requirements, consideration must be given to the severity of service, the reliability expected, and the service life anticipated.

Next to the matching of the vehicle to desired snow and ice control components, creating a vehicle package which is easy to service and maintain makes the most significant contribution to reliability and "uptime" during the storm. Mechanics and service personnel need a complete understanding of the vehicle, equipment design, and requirements. Documentation of service requirements for all items, not just the chassis, should be required of suppliers. A parts list and ready access to all parts is equally important. Preventive maintenance programs based on the manufacturer's recommendations and the severity of service ensure not only reliability but increased longevity.

High-grade paints and coatings, rustproofing packages, and installation of easy-to-clean components should be considered whenever the vehicle will carry snow and ice control chemicals and is expected to have an extended life cycle. Unlike typical highway-use vehicles, special-use snow and ice vehicles frequently reach the end of their service life as a result of corrosion of structural body parts well before the mechanical components fail.

9.7 Alternative Prime Movers

After the motor truck, in its variety of configurations, the most commonly used prime mover for snowplow equipment is the motor grader. Since this machine is designed as an earthmoving device, it has the advantage of being a moderately effective snow removal tool in its basic form. Because of their ability to apply downward pressure on the cutting edge of the standard moldboard, graders are frequently used to cut or scrape packed snow and ice from surfaces. Like a plow, a grader can roll material to the side of the road. As a critical maintenance tool for unimproved and aggregate-surface roads, the grader is often the first available choice for snow and ice control for those maintaining secondary and low-volume road systems. Adding a traditional front snowplow, either blade or V-type, allows the grader to function the same way as a similarly equipped truck. Lacking only the ability to travel at true highway speeds, the grader can in many cases be equally effective. Addition of side wing plow(s) further adds to versatility in high-volume snow regions. A well-equipped operator's cab with heater/defroster is a must. Optional driving front axles, creating a 6×6 machine, are worthy of consideration when large front plows are used. Often the only significant drawback to motor graders is their inability to transport and apply deicing chemicals in addition to plowing.

Front-end loaders are also used for certain snow and ice operations. Typically used to load trucks with chemicals or abrasives, loaders can be equipped with a front plow in place of the bucket for plowing operations in confined areas where the articulated steering of the loader increases maneuverability, or they may be used with the standard bucket attachment to open roads closed by large volumes of drifted snow. This approach is seldom as efficient as large V-type plows or blowers, but it can be effective when such conditions occur infrequently. Loaders can also be equipped with blade and wing plows to deliver performance nearly equal to that of a grader. This, however, is not nearly as common as equipping a loader with an independently powered blower attachment. The loader's four-wheel drive, articulated steering, and bucket arm reach make its application as a blower carrier very practical. The loader also has the added advantage of being a multipurpose machine instead of a dedicated blower carrier.

Many specialized "carriers" for custom snow removal applications are available. These are most common in airport applications, but they can serve highway needs when severe conditions warrant. One hybrid vehicle which cuts across the lines separating truck, special carrier, and tractor is the Mercedes Benz Unimog. Commonly used in Europe and to a much lesser extent in the United States and Canada, the machine resembles a short-wheelbase 4×4 truck. Utilizing larger tractorlike tires and a heavily geared transmission for slow-speed and on-highway operation, the carrier is designed to accept a variety of different tools, including both blade and rotary snowplows. With a very high level of versatility, the Unimog perhaps most closely fits the description of a true multipurpose support vehicle. Unfortunately, the cost and limited availability of specialty carriers in general often makes other solutions more desirable.

9.8 Conclusions

Just as snow and ice conditions vary widely throughout the world, the equipment needed to meet management and removal requirements varies more than ever before. In selecting a prime mover for snow and ice control equipment, it is vitally important to understand the snow conditions, the frequency of storms, the level of removal required, and the equipment necessary to meet the task. To this must be added the limitations on purchase and operating funds. Modern motor truck manufacturing practices offer tremendous ability to customize a vehicle to meet many diverse needs, including many of the requirements of snow and ice control. Add the aftermarket suppliers' and allied equipment installers' capabilities to further refine the product and the ability to meet nearly any special need is almost endless. The keys are understanding the application, understanding the equipment and vehicle options, and exercising the persistence to marry the pieces of the puzzle together into a productive, efficient, and effective tool.

9.9 Specification Checklists

The huge number of options manufacturers offer for their trucks can result in a purchaser's overlooking some components that will provide greater productivity, reliability, and maintainability. As an aid in interpreting manufacturers' literature, Fig. 9-3 identifies customary codes for important dimensions. Checklists should be used by truck purchasers as a guide in preparing specifications for necessary equipment (see Fig. 9-4). By conveying these specifications to the intended truck supplier, the buyer can ensure that the agency's needs are met. The checklists shown contain *suggested* components that will need to be specified.

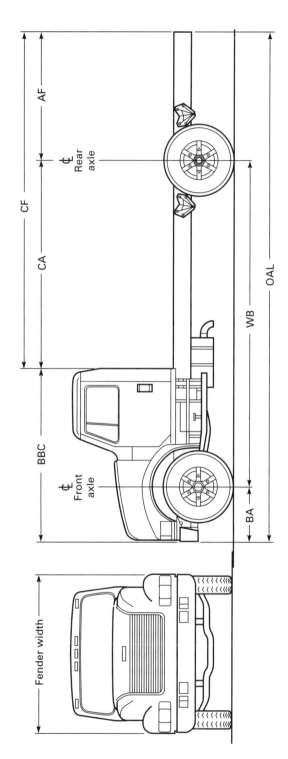

Abbreviations:
AF = Distance from the centerline of the rear axle to the end of the frame
BA = Distance from the bumper to the centerline of the front axle
BBC = Distance from the bumper to the back of the cab
CA = Distance from the back of the cab to the centerline of the rear axle or bogie
CF = Distance from the back of the cab to the end of the frame
OAL = Overall length, the distance from the bumper to the end of the frame
WB = Wheelbase, the distance between the centerline of the front axle and the centerline of the rear axle or bogie

Figure 9-3. Basic truck design showing common dimension codes: (a) two-axle type.

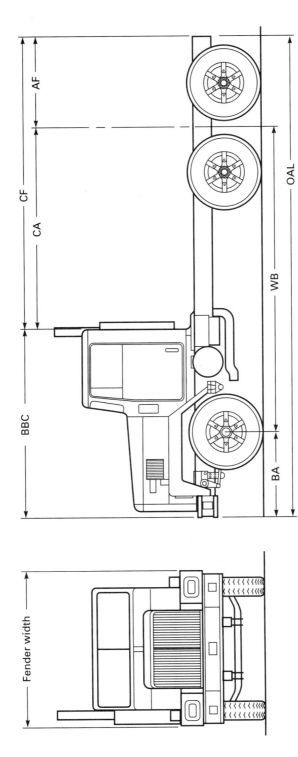

Figure 9-3. (*Continued*) Basic truck design showing common dimension codes: (*b*) three-axle type.

Snow Plow Truck Specification Check List

Chassis:

Capacity:

Total GVW Rating:_____
G.A.W.R. - Front: _____
G.A.W.R. - Rear: _____

Cab-to-Center of Rear Axle Dimension:

Cab-To-Axle: _____Length

Frame:

Frame section Modulus: _____
RBM Rating: _____
Reinforcement -Description:

Front Axle:

Capacity: _____

Front Suspension:

Capacity: _____

Brakes:

Air or Hydraulic?

Type: _____

Brake Accessories:

Automatic reservoir drain valve
Automatic slack adjusters, front and rear
Air Dryer

Emergency Brake:

Horn:

Electric or Air?: _____

Power Steering:

Exhaust System:

Muffler, Horizontal or Vertical Exhaust Pipe and Heat Guard.

Electrical System:

Alternator:

Rated Output: _____Amps

Battery:

CCA at 0° F.: _____

Electrical Circuit Protection:

Circuit breakers
Rating

Headlights With High/Low Beams:

Type
Lens color

Combination Tail/Stop and Rear Signal:

Type

Front Turn Signal and Parking Lights:

Backup Light:

Cab Clearance Lights:

Cab-Mounted Identification Lights:

Rear Identification Lights (3-Bar):

Body Clearance Lights:

Reflectors

4-Way Hazard Switch:

Engine Accessories:

Coolant Filter
Air Filter
fuel moisture separator
fuel heater
Cold Weather Starting Aid
In-Block Type Engine Heater

Engine:

Make: _____
Model: _____
RATED H.P. _____ @ _____ RPM

Figure 9-4. Specifications checklists.

Trucks for Snow and Ice Control

Transmission:

Make: _____
Model: _____

P.T.O. Opening:

Rear Axle & Rear Suspension:

Capacity: _____
Ratio: _____

Rear Suspension:

Capacity: _____

Fuel Tanks:

Total Tank Capacity: _____

Front Wheels & Front Tires:

Type: _____
Size: _____
Oil type seals.

Front Tires:

Type _____ :
Size Tire: _____

Rear Wheels & Rear Tires:

Type: _____
Size: _____

Rear Tires:

Type: _____
Size Tire: _____

Cab:

Minimum cab dimensions:
Shoulder Room: _____
Floor to Ceiling: _____
Convenience Items:
    12 Volt Socket W/Plug
    Dome light.
    Dual arm rests.
    Dual outside grab handles.
    Dual sun visors.

Front Fenders & Hood Assembly:

Material: _____

Note: Access shall be provided for routine maintenance without removal of the hood assembly or the snowplow push frame.

Front Wheel Splash Guards:

Paint:

Color:
Areas Painted:

Rustproofing:

Type
Areas of truck to be covered:

Seats, Seat Belts, & Reels:

Instruments and Warning Systems:

    Speedometer
    Odometer
    Tachometer
    Fuel gauge - indicating all tanks
    Hour meter
    Engine-coolant temperature gauge with high temperature warning light and buzzer
    Air pressure gauge with low-pressure warning light and buzzer
    Engine oil pressure gauge with low-pressure warning light and buzzer
    Hand throttle
    Engine air filter restriction gauge

Rear View Mirrors:

Type - Dual, electrically heated, outside, rear vision.

Heater & Defroster:

Heavy-duty, hot water, fresh air type with all controls for effective heating and/or defrosting under cold weather conditions.

Windshield Wipers:

Type - Dual, heavy-duty, electric, two-speed windshield wipers with intermittent speed control.
Blades - Rubber covered

Figure 9-4. Specifications checklists (*Continued*)

Special Truck Wiring:

General Requirements:

All wiring terminal connections shall be crimped and soldered. All wiring must be uninterrupted and completed without splices of any kind.

Auxiliary Snow Removal Headlights with Park/Turn Lamps:

Warning Lights:

Additional Lights:

Body Height Indicator:

Backup Alarm:

Type
Sound level

HYDRAULIC SYSTEMS:

Hydraulic System for Dump Trucks/Snowplows:

A complete hydraulic system for operation of dump body hoist, hydraulically powered ice control spreaders, and hydraulically operated snowplows.

Hydraulic Lines and Piping:

Ice Control Spreader System:

Hydraulic Pump Drive:

Hydraulic Pump:

Oil Reservoir:

Capacity: _____

Valve Bank:

Pressure Relief Valve:

Spreader Control:

Truck Dump Body:
General:
Gauge: _____
Yield strength: _____
PSI rating: _____
Inside length: _____
Inside width: _____
Overall width: _____
Water level capacity: _____

Tailgate:

Height
Controls

Rear Wheel Mud Shields:

Rear Wheel Splash Guards:

Cab Shield:

Body Hoist:

Type
Capacity
Dump angle: _____ °

Safety Prop:

Snowplow Push Frame:
General:

Type
Push lug spacing
Connecting pin holes
Push lug height adjustment

Construction and Design:

Hydraulic Lift Arm:

Snowplow Lift

Bumper

Complete Truck and Component Equipment Warranty Coverage:

Delivery:

Operator's Books:

Service Manuals:

Training:

Figure 9-4. Specifications checklists. (*Continued*)

10
Railroad Snow and Ice Control

10.1 Introduction

Today we have no conception of and little appreciation for the difficulties our forebears had in traveling, particularly in winter. Though intercity travel by wheeled vehicles such as stagecoaches was facilitated when the unpaved roads froze and wheels no longer had to push through mud, snow was often a barrier. This was overcome by mounting the vehicles on runners. Deep snow could impede even sleighs and the horses drawing them, so snow rollers were frequently used to compact the accumulation to improve passage. The introduction of railroads changed this picture dramatically, and led directly to the development of snow removal equipment. The earliest railroads, used in mining in the sixteenth century, used carts pulled by horses, running first on wooden poles, then on cast iron rails. It wasn't until 1804 that a locomotive was built in England that was capable of pulling heavy loads, and not until 1825 that steam-powered passenger and freight service began in England. The first railroad in the United States, in Quincy, Massachusetts, was built in 1826 and used horse-drawn wagons to haul minerals. Passenger service was introduced in 1828 when the Baltimore & Ohio Rail Road began operation using horse-drawn cars, and steam power was introduced on the B&O in 1830. It was recorded in a scientific journal in 1831 that the B&O successfully cleared the tracks for its trains by sending a team of horses ahead with a "contrivance" to push the snow aside (Sullivan 1831). Deep snows were frequently removed by shovels wielded by laborers. Newspapers recorded the delays that snow caused. On January 5, 1859, the *New York Times* reported on its first page that "The Long Island Railroad Company were hard at work all day, but did not succeed in getting their trains into operation." And in a column datelined "Springfield, Mass., Jan. 4—5 P.M. The Express train, which left Boston early this morning for New York, and due here at fifty minutes past eleven is fast in the snow, two miles east of Worcester, and nearly covered up....The snow is two feet deep, and still falling" (*New York Times*, Jan. 5, 1859, p. 1).

Necessity is the mother of invention, and the imagination of inventors was stimulated. Records show that two patents were awarded for snowplows in 1846, and following big storms in 1856 a dozen more submitted designs (Ludlum 1982). McKelvey (1995) reported that

> The New York Central brought an 'improved snowplow' to Rochester [NY] in January 1857 for use on its upstate lines. The new plow, invented in Philadelphia, was mounted on wheels and had a broad shovel scoop to lift the snow from the tracks, and a wedge near the back or top of the tilted scoop to shove the snow off to the side, or onto a movable chain designed to carry it back to a dump car behind.

The westward expansion of railroads in North America led across the Sierra Nevada and Rocky Mountains with their heavy and frequent snowfalls, avalanches, and deep wind drifts. The Central Pacific Railroad began construction in Sacramento, California, and crossed the Sierra Nevada in California, meeting the westward progress of the Union Pacific at Promontory Point, Utah, in May 1869. The lightweight snowplows that up to that time had been adequate to cope with eastern snows proved wholly inadequate for the western conditions. These were pilot plows, small wedge-shaped sheet-iron contrivances attached to the pilot (commonly called the "cowcatcher" in the United States). In the moderate snowfalls of the east, frequent train operations were sufficient to keep the tracks clear and open. Heavy snows and snowslides in the west made this method of keeping a line open inadequate. During the first winter of construction in 1865–66, thousands of laborers using hand shovels were employed to keep the completed sections of the railroad clear of snow. This prompted the fabrication of a heavy wedge-shaped or "bucker" plow in the Sacramento shops of the Central Pacific in 1866. This was the forerunner of the wedge plow used to this day (see Fig. 10-1). The first unit weighed 12 tons (11,000 kg) and was mounted on two standard freight car trucks. The wedge of iron plates flared up at an angle of 45°, then flared outward. In the first winter of use, three 36-ton (33,000 kg) locomotives pushed the

Figure 10-1. Russell snowplow.

plow into the deep snow at top speed until it was brought to a stop. It was then backed out and once again run at full speed into the snow. If the snow was deeper than the plow and it became stuck, the track under the plow's trucks had to be shoveled clear of snow to prevent derailment. The weight of the plow was insufficient to prevent snow from building up under the bottom of the plow, a situation that caused it to derail. A sturdier second plow was fitted with pig iron to increase the weight to 19 tons (17,000 kg). It held the rails. As many as 12 locomotives were necessary to push this heavier plow through deep snow (Best 1966).

The inadequacy of the bucker plow in pushing aside deep snow led to the invention of the rotary plow. Perhaps reflecting his experience as a dentist, J. W. Elliot of Toronto, Canada, was the first to conceive the rotary plow principle. His "Revolving Snow Shovel" was patented in 1869. He built a small hand-operated model but did nothing more to promote his idea. In the same year, Charles W. Tierney of Altoona, Pennsylvania, received a patent for a two-stage rotary plow consisting of a revolving screw that fed snow to a fan behind it. This also progressed no further than a scale model. The first rotary plow to be built and actually used was the Hawley plow, which was displayed at the 1876 Centennial Exposition in Philadelphia. It incorporated a large vertical screw that was fed snow through a cylindrical casing in front. It was a total failure (Best 1966). The first successful rotary plow was designed by Orange Jull of Orangeville, Ontario, the owner of a flour mill, and built by John S. and Edward Leslie in their machine shop in Orangeville. The design incorporated a rotating set of cutters in front of the plow that fed snow to a fan behind (Winterrowd 1920). Improvements in the basic design of these so-called Leslie plows were made over the years; nearly 150 units have been built, and they were once used by most railroads in North America and several foreign countries. Because of the newer snow clearance and snow control procedures described below, the use of these very expensive machines has declined, and today most have been scrapped, are in storage, or are in museums.

10.2 Current Snow Clearance and Snow Control Methods

Appropriate techniques for snow clearance and control are based on the location of the track and other facilities:

1. Main line
2. Yards and terminals
3. Switches
4. Grade crossings
5. Tunnels and snow sheds

10.2.1 Main Line

Railroad snow removal differs from highway snow removal in two important respects: The linear arrangement of the main line and general absence of interfering traffic allows high-speed operation of plows, and tremendous power is available to push through snow accumulations at a range of speeds to disperse snow well away from the track. Frequent passes of small pilot plows may be sufficient to keep the track clear along level, unobstructed sections at grade or higher, but heavier equipment may be necessary where deeper snows have accumulated in cuts. On-track equipment used for the purpose includes wedge plows, snow dozers, and flangers (Sec. 10.3.1). Several practices have reduced the need for heavy snow removal equipment in recent years. The method that has the greatest long-term benefits involves flattening the slopes on either side of the right-of-way to reduce accumulation of wind-blown snow. When snow does build up alongside the right-of-way, bulldozers are frequently used to flatten the accumulations. This has proved to be a cost-effective method. Bulldozers or front-end loaders are also used to clear deep accumulations on the track when a rotary plow is not available or it would be too costly to bring one in.

10.2.2 Yards and Terminals

Platforms; mail, cargo, and industrial areas; walkways; roads; switching leads; and classification facilities generally must be cleared of snow. Because of the congestion that exists and the lack of storage space for snow cleared from these facilities and from tracks, snow must frequently be loaded on trucks or railcars and hauled to disposal areas. Small walk-behind snowblowers similar to those used by homeowners are commonly used to clear station platforms and other limited areas, as are power rotary brooms mounted on small tractors. Hand shoveling is used only when mechanized equipment is not available or space is too limited for maneuverability.

10.2.3 Switches

Cleaning snow and ice from switches (also called points or turnouts) is one of the costliest and most critical winter maintenance tasks faced by a railroad. Conventional moving-point switches will not function properly when snow collects between the moving and stationary elements and becomes compacted. This is critical in the case of remotely operated switches, particularly those located on automatically controlled centralized traffic control (CTC) sections. The point rails of a turnout swing laterally on slide plates that are machined to fit closely to the head and base of the stock rails. As little as ¼ in (0.6 cm) of snow that becomes compacted may result in a gap between the fixed and moving rails and possibly cause a derailment (Ringer 1979). In the early days of railroading, clearing snow from switches was accomplished entirely by men using brooms and shovels. Though this manual method is still used at times in yards, built-up areas, and other selected locations, equipment has been developed to reduce the labor cost and to enable clearance in remote locations. When the equipment operates automatically, no extra labor is required.

Two classes of devices are used: fixed installations and mobile equipment. Switch heaters and blowers are used for fixed installations, and mobile equipment includes blowers (either heated from a combustion jet or only a cold air blast) and brooms. Snow control methods are classified as follows:

1. Fixed installations
   a. Heaters
      (1) Open-flame
      (2) Contact
         (a) Electric
         (b) Fuel-fired
         (c) Steam
      (3) Forced convection
      (4) Jet (pulse)
   b. Air curtain (blowers)
2. Mobile equipment
   a. Blowers
   b. Air jets
   c. Brooms

Many methods of applying heat to switches have been tried over the years, beginning as early as the 1890s. Thermal protection methods have the advantage of being able to clear a switch that has been completely covered by hard snow that cannot be removed by nonthermal methods. Directing an open flame fueled by propane, fuel oil, kerosene, or natural gas onto the rails is a common method. The kerosene pot burner was one of the earliest types of switch heaters. This has been replaced for the most part by more efficient designs, but it is still used as an expedient in locations not normally protected by permanent heaters. Modern heaters of this type can be ignited manually where workers are customarily present, such as in yards, but automatic igniters are used for remote locations.

Precautions must be taken when using thermal systems to prevent failure:

1. Melted snow must be drained away from the switch being protected to prevent refreezing and buildup of ice, which could cause derailment.
2. Both open-flame and hot-air heaters may burn wooden ties; installation of asbestos sheets on the sides of ties can reduce this probability.
3. Frozen ballast may be thawed by heaters and result in settlement of the rail, which may require shimming to maintain alignment.

Blowers, pulse jets, and air curtains provide a high-volume blast of air to the moving switch parts to prevent snow accumulation. The blower or compressor unit is located near the switch, and ducts deliver the air to the points. Generally the air is heated, using natural gas or propane as fuel.

Mobile equipment includes blowers or jet engines mounted on flat cars which direct a high-velocity blast onto the switch, compressed air jets, and power brooms on specialized track-cleaning maintenance of way (M.O.W.) cars.

10.2.4 Highway Grade Crossings

Flangeways must be kept clear of compacted snow and ice to reduce the likelihood that railcar wheels will be lifted and cause derailments. Manual clearing using brooms, picks, and shovels is one method that is still used, but power rotary brooms and compressed air jets have largely replaced hand methods. Ice-melting chemicals are also used to melt ice in the flangeways. Another method that is sometimes used is to fill the gap with either solid elastomeric material or sealants. Wheels rolling across crossings protected in this manner may displace the materials. These can be recovered in the spring and reused. In rural and agricultural areas, temporary crossings consisting of planks inserted between the rails and on either side are removed as winter approaches so that mechanical snow removal equipment can reach below rail height.

10.2.5 Tunnels and Snow Sheds

Water often seeps naturally into tunnels from overlying rock and soil strata. Cold winds blowing through the tunnels can result in ice forming on the walls or on the track, causing problems in both locations. Wall and roof formations can damage rolling stock and power and communications cables, and even the tunnel lining itself. Ice accumulating on or between the rails can cause derailments. Ice also becomes a threat to the safety of maintenance and operating personnel. The frequency and degree of ice formation will depend on the number of freeze–thaw cycles, the availability of water to contribute to the seepage, the permeability of soil and rock formations surrounding the tunnel, and the wind direction and velocity with respect to the orientation of the tunnel openings.

Infrequent or minor ice accumulations can be removed manually by chipping, or icicles can be dislodged by equipment passing through. Ice that is dislodged must be hauled out of the tunnel. Lining the tunnel with insulation has been successful in preventing ice formation in the Belden Tunnel near Binghamton, New York, and in the Hoosac Tunnel near North Adams, Massachusetts (Mudholkar 1991). Note, though, that these tunnels are located in a relatively mild climatic region.

Snow sheds are structures built in mountainous areas where snowslides occur year after year, making construction of permanent protection economically justified. These are usually open on the downslope side to reduce the cost of construction and to facilitate removal of any snow which may blow in or be carried in by trains. They were first used in the western United States during the construction of the transcontinental railroad across the Sierra Nevada, where heavy snowfalls overwhelmed the plows then available as well as the armies of track workers wielding shovels. Many miles of snow galleries, or man-made tunnels, were constructed. Wood was used for constructing the first snow sheds, many of which burned down when ignited by sparks from coal- or wood-burning locomotives. Though that danger has passed with the ascendancy of diesel locomotives, most modern snow sheds are constructed of reinforced concrete. Few snow sheds remain in North America, though their use in Switzerland has not declined.

10.3 Snow Removal Equipment
10.3.1 On-track

Pilot plows, small V-shaped steel attachments on the front of locomotives, are sufficient for removing small accumulations of snow. Frequent train movements will clear snow even in fairly heavy snowfalls. This will result in clearance only to track height. When snow does accumulate, either because of very heavy snowfall or on lines where traffic is less frequent, specialized equipment is used. Flanger cars have been used for many years to remove snow between the rails. A flanger is a steel blade that is dropped below rail level between the rails and on the side of each rail. An operator controls the height of the blade and raises it to avoid striking turnouts and crossovers. The use of flangers is declining; some roads have ended their use entirely. Heavy snows are cleared with wedge plows. Two types of wedge plows are used: the single-track plow in the form of a V that displaces snow to both sides of the track, and the double-track wedge plow that moves snow to one side to avoid covering an adjacent track. The latter plow is similar in function to a one-way highway plow. In the early V-shaped designs, the plow could be shifted to right or left to serve the same purpose of moving snow to only one side.

Rotary plows have largely been replaced by other equipment, although a few units remain on standby on western U.S. roads. The Leslie plow described above was the principal design for many years, and several were exported to Europe. Newer, more efficient designs have been introduced in Europe, including the Beilhack and Rolba, both of which are two-stage augur units (described more fully in Chap. 5). Rotary plows were originally steam-powered, but they are now powered by electric motors. The plow units are not self-powered, but are typically operated with two diesel engines, one to provide electric power for the plow and the second for propulsion.

A unit called a spreader has taken over much of the clearance task, replacing the flanger and wedge plow in other than deep snow (Fig. 10-2). Though originally designed for roadbed maintenance, it has proved valuable for snow clearance. This is familiarly known as a Jordan spreader after its manufacturer. It has movable wings on both sides that can be raised and lowered or moved in and out by hydraulic actuators to change the width of the cut. It is not self-propelled.

10.3.2 Off-track

Because of the difficulty in traversing rough terrain to reach track that is covered with snow, tracked vehicles are necessary. Bulldozers are most frequently used, although it often becomes necessary to employ any excavating equipment that can reach the snow problem area. Gradalls, rotary plows, and highway blade plows are used where possible, when the ground is frozen hard enough to support their weight.

Figure 10-2. Jordan spreader.

10.4 Electric Systems

Many rail networks in winter climates, particularly in the eastern United States and in Europe, Scandinavia, and Japan, have extensive electrified sections. Transit systems—that is, heavy or light rail systems operating in urban areas—are also largely electrified. These routes have many of the problems caused by snow already described plus some that are unique to their systems. Though these systems are outside the scope of this book, brief mention will be made of the most significant of these problems. Electricity for the drive motors is obtained from either conductor rails (referred to as the third rail in the United States) or overhead catenaries. Ice buildup can prevent shoes or pantographs from making contact with the power source, and various methods to combat this have been investigated.

11
Airport Snow and Ice Control

11.1 Introduction

Many of the snow and ice control procedures used at airports and on highways are similar; the objective in both environments, of course, is to reduce interruptions to normal service and to provide safe conditions for vehicle movements. Airports actually have two environments, the airside (runways, ramps or aprons, and taxiways—all the surfaced areas over which aircraft will pass) and the landside or groundside (the road system providing access from the public highway to the terminals). Equipment used for winter maintenance of the airport landside facilities is indistinguishable from that used on public highways. There are, however, a number of differences in the types of chemicals, equipment, and techniques used on the airside.

Chief among the differences between airport and highway is the matter of time. Delays in aircraft takeoffs and landings mean lost revenue to the carriers and to the airport authorities. Costs reported in 1992 for carrier delays amounted to $11/min for ramp delays, $21/min for taxiing delays, and $43/min for delays while airborne. The annual cost to all U.S. carriers amounted to $8 billion. What more incentive is needed to incur extra effort and expense to reduce delays? This is reflected in larger plowing equipment, use of specialized equipment such as power brooms with high-velocity air blast and high-capacity chemical and abrasive spreaders that can serve wide areas with a minimum of passes, and the need to dispose of snow from the congested aprons where storage space is insufficient. Another difference is the restriction on the use of ice control chemicals that could damage the highly stressed metals on aircraft. This rules out the chlorides and increases the dependency on other chemicals, which in the past have been less effective in their action. This in turn has led to some dependence on sand for friction improvement on icy runways, even though there is some risk of damage to jet engines and other aircraft components. Operations unique to the airport airside are described in the paragraphs that follow. Portions are based on the Federal Aviation Administration Advisory Circular *Airport Winter Safety and Operations* (AC 150/5200-30A, dated 10/1/91, including change 2 dated 3/27/95). Printed copies can be obtained from the Federal Aviation Administration or may be downloaded from the World Wide Web at <http://www.faa.gov/arp/150acs.htm>.

11.2 Equipment

Characteristics of common and specialized airport equipment have been included in the discussion of snow-removal equipment in Chap. 5. Operational aspects of the equipment are described in more detail here.

11.2.1 Blade Plows

As for highways, greatest dependence is placed on displacement, or blade, plows for removing the bulk of snow accumulations and attempting to remove ice if it has been allowed to form. Wind direction is frequently a factor in plowing operations, and unlike the case with highways, a choice usually can be made to plow to either side of a runway depending on the wind direction. Whether one plow is given the task of clearing a runway or, as is more frequently the case, two or more plows work in echelon formation, snow should be plowed in the direction of the wind. This will reduce the amount of snow that will blow back onto the cleared area and that may also obstruct the driver's vision. Picture what the plow orientations must be: Let's say that the first pass on a north-south runway is made heading north and plowing to the right, that is, east. On the return pass, the truck is heading south, so in order to continue moving the snow off the runway in the same direction, or to the east, the plow must cast to the left. Clearly, a fixed, one-way blade cannot do this. It's a job for a reversible plow, one that can be shifted to cast in either direction. Power-operated reversible plows controlled from the truck cab are used to speed the work. A unique type of reversible plow used on airport runways and not on highways is the rollover, originally used at U.S. Air Force bases to handle high volumes of snow at relatively high speeds. These plows are heavy and project well in front of the carrier truck. The large truck with heavy-duty front axle required for rollovers requires a wide turning radius, something easily accommodated on a runway but not possible on most roads (see Fig. 5-4).

11.2.2 Brooms

Powered brooms or brushes have proved to be a highly useful tool for primary snow removal and for cleanup. They are capable of removing light accumulations of snow without leaving a residue, whereas plows generally will leave a residue because of pavement irregularities and plow bouncing. Large airports have large brooms equipped with a high-velocity blower that removes any snow, ice, or debris dislodged but left by the broom. Sand applied for friction improvement is readily removed by brooms. The air blast can be used to clear snow from runway edge lights, although well-consolidated or deep snow must be removed by other means, such as shoveling or air jets (see Fig. 5-10).

11.2.3 Rotary Plows (Snowblowers)

Snow moved to the edges of runways, taxiways, or aprons by displacement plows may form snowbanks which are high enough to be struck by aircraft wings. The height must be reduced to provide wing overhang clearance and engine clearance when the airplane's wheels are on the full-strength pavement. Rotary plows are most frequently used for this purpose. They can cast the windrowed snow beyond the edge of the pavement and disperse it over a wide area. Small rotaries can move 1000 tons/h (907 t/h); units are made with capacities to 10,000 tons/h (9070 t/h).

11.2.4 Spreaders

The extensive areas of runways and taxiways require large-capacity spreaders capable of distributing abrasives or solid chemicals over a wide swath. Twin-spinner spreaders have been designed for airfield use. The width of spread can be varied by modulating the spinner rotational speed, or, with some designs, the entire spinner assembly can be tilted to provide a higher particle trajectory (Fig. 11-1).

Figure 11-1. Airport spreader with dual spinners that can be tilted to control spread width.

11.2.5 Liquid Applicators

The necessity to apply liquid chemicals on wide runways has given rise to spray trucks fitted with booms extending up to 24 ft (7.3 m). The booms are hinged so that they can be folded to no more than the width of the vehicle to allow parking in garages and other restricted areas.

11.3 Chemical Uses

Airports use ice control chemicals on the airside for two principal purposes: to eliminate snow and ice on aircraft movement surfaces, thereby providing a safe surface, and for deicing aircraft prior to takeoff. This manual will not cover the techniques of aircraft deicing. However, the glycol-based chemicals used for that purpose differ from the chemicals used for treatment of pavement, so they will be described here. Much of the chemical applied to aircraft falls onto the pavement during taxiing and takeoff. It is estimated that 20 percent of the glycol used adheres to the aircraft and 80 percent runs off to the deicing pad; 80 percent of the glycol adhering to the aircraft will blow off onto the runways/taxiways during taxiing and takeoff (BWI 1993).

Airport Snow and Ice Control

11.3.1 Aircraft Deicing/Anti-icing Fluids

Aircraft deicing fluids (ADF) are applied to wings, elevators, and rudders to remove ice and to prevent ice from forming until the aircraft is airborne. There are two types of ADF, Type I and Type II. Both types use propylene glycol or ethylene glycol as their base. Type II fluids include a thickening agent that usually enables them to adhere to the wings. The lower-viscosity Type I fluids will flow off the aircraft rapidly and remain effective for a much shorter time, and so they must be applied immediately before departure. Type II fluids have a longer holdover time, about 30 to 35 min, then will progressively flow off the wings as the aircraft reaches takeoff speed. Either type may have to be reapplied if the aircraft sits on the ground for even a few minutes. Wider use of Type II fluids is now required by the Federal Aviation Administration. Only ethylene or propylene glycol–based aircraft deicing fluids are certified by the Society of Automotive Engineers (SAE) and approved by the FAA.

11.3.2 Chemical Control on Runways, Taxiways, and Ramps

The prohibition against using corrosive chemicals for control of snow and ice because of the potential damage to aircraft metals has limited the choice to urea and ethylene glycol until the last few years. Both of these chemicals are now considered to damage the environment, urea because of its high nitrogen content and high BOD, and ethylene glycol because of both its high BOD and its toxicity. Their use is no longer permitted at most airports. Ethylene glycol-based fluids continue in use for deicing aircraft.

11.3.3 Liquid Chemical Application Rates

Determining the variables of equipment and chemical concentration that affect the recommended or desired amount of liquid chemical to apply can be confusing. A common application rate for airport runways is given as so many gallons per 1000 ft^2 (liters per 100 m^2). What flow rate is required to provide this with your system? A little bit of mathematics will tell us.

English Units

$$F = (SWG)/11.36D$$

where F = flow rate, gal/min
W = spray bar width, ft
G = application rate, gal/1000 ft^2
D = dilution factor
S = speed of spray apparatus, mi/h

Example. The recommended application rate for the ice control chemical you want to apply based on site conditions (temperature and temperature trend, pavement type, whether deicing or anti-icing, chemical response time, traffic) is 0.2 gal/1000 ft^2 with the full-strength chemical ($D = 1$). With a width of spray of 6 ft, and applicator vehicle speed of 25 mi/h, the flow rate is nearly 3 gal/min.

$$F = (25)(6)(0.2)/11.36 = 2.6 \text{ gal/min}$$

If the chemical had been diluted 1 part full-strength with 2 parts water, the dilution factor would be $\frac{1}{3}$ and the equation would read $F = (25)(6)(0.2)/(11.36)(1/3) = 7.9$ gal/min. The most probable use of dilution would occur when the speed of the application vehicle would be too fast for the specified application rate (which for reasons of compatibility with the spray apparatus you wish to maintain at a certain value). The equation can, of course. can be rearranged to solve for any of the variables.

Metric Units

$$F = (SWG)/6D$$

where F = flow rate, L/min (L is abbreviation for liters)
W = spray bar width, m
G = application rate, L/100 m^2
S = speed of spray apparatus, km/h
D = dilution factor

Example. The recommended application rate for the ice control chemical you want to apply based on site conditions (temperature and temperature trend, pavement type, whether deicing or anti-icing, chemical response time, traffic) is 0.82 L/100 m^2 with the full-strength chemical ($D = 1$). With a spray width of 1.8 m, and applicator vehicle speed of 40 km/h, the flow rate F is nearly 10 L/min.

$$F = (40)(1.8)(0.82)/6 = 9.8 \text{ L/min}$$

If the chemical had been diluted 1 part full-strength with 2 parts water, the dilution factor would be $\frac{1}{3}$ and the equation would read $F =$

(40)(1.8)(0.82)/(6)(1/3) = 29.5 L/min. The most probable use of dilution would occur when the speed of the application vehicle was too fast for the specified application rate (which, for reasons of compatibility with the spray apparatus, you wish to maintain at a certain value). The equation, of course, can be rearranged to solve for any of the variables.

11.4 Friction Testing and Reporting

The principles of friction and its measurement are given in Chapter 4. Certain aspects of friction measurement are specific to airfield operations; they are described below.

Measurement of the frictional characteristics of a runway surface is an accepted method for determining an aircraft's braking performance at the speeds experienced during takeoff and landing. Decelerometers (DEC) were the first devices used for determination of braking action on snow- or ice-covered runways, beginning in the early 1950s. These devices require hard braking of the test vehicle, and thus only spot measurements are produced. A second type was developed later: continuous-friction measuring equipment (CFME). As the name suggests, these produce a continuous-friction record. The James Brake Decelerometer was the first device commonly used, giving rise to a James Brake Index (JBI) relating the readings to friction and expressed in the form of nomographs for determining corrected landing distances. The United States Air Force developed similar charts based on James Brake Declerometer readings, and referred to them as Runway Condition Reports, or RCR. Decelerometers available currently are either mechanical or electronic; the former require operator calculation of the friction value; the latter process the measurement data electronically and display a value. Electronic units are rapidly replacing the mechanical, since measurements can be made more rapidly and time on the runway between flight operations is usually severely limited. Decelerometers are now available from several manufacturers. Continuous-friction measurement devices are available in two designs: self-contained in the vehicle, and as a towed unit. Only equipment which has been certified by the FAA may be used on commercial airfields in the United States. A current list of approved equipment can be found in Appendix 4 of AC 150/5320-12C (FAA 1997). The two types of friction measuring equipment, DEC and CFME, give equivalent results for friction values of 0.4 and below (in the United States, friction values are frequently expressed as "MU values," using a multiplier of 100 to produce whole numbers; e.g., a MU value of 45 would equate to a coefficient of friction of 0.45). It has been found by repeated testing that speed of the test vehicle does not have a significant influence on the friction measurement made on compacted snow or ice; low friction values on these surfaces can occur at low speeds [below 20 knots (32 mi/h)] and at high speeds [greater than 100 knots (115 mi/h)] (Anon. 1995).

The International Civil Aviation Organization (ICAO) has published a table of estimated aircraft braking action for a range of friction coefficients and assigned descriptors to each range (Table 11-1). These values are not absolute and are to be considered only as relative values; "good," for example, does not mean that a pavement condition is that of a dry, clean runway but only that an airplane should not experience directional control or braking difficulties when landing. All reports of friction values are advisory, since they are qualitative and transitory in nature.

ICAO recommends that runway surface friction be reported for each ⅓ segment of a runway. By international agreement, these are designated A, B, and C. Section A is the lower runway designation number. Information given to a pilot before landing refers to the first, second, or third part of a runway, where the first ⅓ segment is where the aircraft will touch down. ICAO recommends that friction measurements be made in two parallel tracks, each approximately 10 ft (3 m) on each side of the centerline, or the distance of the main landing gear of most aircraft using the airport. The average friction value should be determined for each segment A, B, and C.

The Federal Aviation Administration supplements the ICAO recommendations by stating that on runways serving primarily widebody aircraft, friction surveys should be conducted approximately 20 ft (6 m) from the runway centerline. Friction measuring equipment should be operated in the same direction that aircraft are landing. A runway is divided into three equal zones: touchdown, midpoint, and rollout, defined according to aircraft landing direction. FAA recommends a minimum of three braking tests in each zone, to be made in one con-

Table 11-1. Aircraft Braking Performance on Compacted Snow- or Ice-Covered Runway (ICAO 1995)

Coefficient of friction	Estimated braking action for aircraft
0.40 and above	Good
0.39 − 0.36	Medium to good
0.35 − 0.30	Medium
0.29 − 0.26	Medium to poor
0.25 and below	Poor

tinuous pass, if possible. If traffic prevents this, each zone must be scheduled separately until all three zones are tested (FAA 1991).

The condition of a runway surface is communicated to pilots in the airport operator's report, the Notice to Airmen, or NOTAM. The following information is included:

- surface condition: wet snow, dry snow, slush
- depth of contaminant
- partial/full coverage of contaminant
- snow bank exceeding heights agreed to in the airport operations plan
- runway braking action reports
- chemical or abrasive treatments
- proposed runway closing time and duration of closure
- obscuration of any centerline, touchdown zone, or edge lights
- marking/signs
- date and time of reading
- pavement temperature and readings of friction measurement equipment, if available
- any other condition that may impair safety

11.5 Snow Desks

A snow control center or snow desk is a special facility set up to serve as the principal source for airfield conditions and to direct snow and ice control operations. It communicates with air carriers and Air Traffic Control to inform them of expected runway closures, and with snow and ice control equipment operators and supervisors to direct operations and to ensure that all equipment has left the runways before air operations are resumed.

11.6 Reference Materials

Federal Aviation Administration (FAA) Advisory Circulars (AC)
150/5200-30A, *Airport Winter Safety and Operations* (10-1-91; change 2, 3-27-95). Provides comprehensive guidance on mechanical and chemical snow removal and ice control, runway friction testing, clearance priorities and times, and pavement condition reporting (FAA 1991).

150/5220-13B, *Runway Surface Condition Sensor Specification Guide* (3-27-91). Provides guidance to assist airport operators, consultants, and design engineers in the preparation of procurement specifications for sensor systems which monitor and report runway surface conditions (FAA 1991a).

150/5220-18, *Buildings for Storage and Maintenance of Airport Snow and Ice Control Equipment and Materials* (10-15-92). Provides guidance for site selection, design, and construction of buildings used to store and maintain airport snow and ice control equipment and materials (FAA 1992).

150/5220-20, *Airport Snow and Ice Control Equipment* (6-30-92). Provides guidance to assist airport operators in the procurement of snow and ice control equipment for airport use (FAA 1992a).

150/5300-13, *Airport Design* (9-29-89; and change 4, 11-10-94). Contains the FAA's standards and recommendations for airport design (FAA 1989).

150/5300-14, *Design of Aircraft Deicing Facilities* (8-23-93). Provides standards, specifications, and guidance for designing aircraft deicing facilities (FAA 1993).

150/5320-12B, *Measurement, Construction, and Maintenance of Skid Resistant Airport Pavement Surfaces* (11-12-91). Contains guidelines and procedures for the design and construction of skid-resistant pavement; pavement evaluation, without or with friction equipment; and maintenance of high-skid-resistance pavements (FAA 1991c).

Society of Automotive Engineers (SAE)

Compound, Solid Deicing/Anti-Icing Runways and Taxiways. AMS 1431A. (SAE 1992).

Fluid, Deicing/Anti-Icing, Runways and Taxiways Glycol Base. AMS 1426C. (SAE 1993).

Fluid, Deicing/Anti-Icing, Runways and Taxiways Potassium Acetate Base. (AMS 1432A. (SAE 1991).

Fluid, Generic, Deicing/Anti-Icing Runways and Taxiways. AMS 1435. (SAE 1995).

Sand, Airport Snow and Ice Control. AMS 1448A. (SAE 1994).

Appendix A

Reference Materials

Calculation of weight percent of a chemical

$$\text{wt\%} = \frac{\text{wt of solute}}{\text{wt of solvent} + \text{wt of solute}} \times 100$$

$$= \frac{\text{wt of solute}}{\text{wt of solution}} \times 100$$

Miscellaneous Useful and Interesting Values and Conversions

Freezing-point depression by pressure: $0.0074°C/10^5$ Pa (14.7 lb/in^2)

mi/h $\times$ 1.60934 = km/h

mi/h $\times$ 0.44704 = m/s

mi/h $\times$ 0.8684 = knots (kn)

kn $\times$ 1.152 = mi/h

1 gal = 3.7854×10^{-3} m$^3 \times 10^3$ L/m^3 = 3.7854 L

1 L = 10^{-3} m^3

1000 L = 1 m^3

Mass (Weight)

lb	oz	kg
1	16*	0.45359
0.06250	1	0.028350
2.20462	35.274	1

1 ton = 907.2 kg = 0.9072 t

1 t = 1000 kg = 2204.6 lb

Volume

Cubic inches (in^3)	Cubic feet (ft^3)	Gallons) (gal)	Liters (L)	Cubic meters (m^3)
1	5.787×10^{-4}	4.329×10^{-3}	0.0163871	1.63871×10^{-5}
1728*	1	7.48055	28.317	0.028317
231*	0.13368	1	3.7854	0.0037854
61.02374	0.035315	0.264173	1	0.001*
61,023.74	35.315	264.173	1000	1

*Values are exact.

Density

lb/ft^3	lb/gal	g/cm^3	kg/m^3 (g/L)
1	0.133680	0.016018	16.018463
7.48055	1	0.119827	119.827
62.4280	8.34538	1	1000
0.0624280	0.008345	0.001	1

Thermal Conductivity

Btu/(h • ft • °F)	cal/(s • cm • °C)	W/(m • K)
1	4.1338×10^{-3}	1.7307
241.91	1	418.68*
0.57779	2.3885×10^{-3}	1

Appendix A

Pressure

lb/in²	in of water (60°F)	in Hg (32°F)	mm Hg (32°F)	Bar	kg$_f$/cm²	Pascal (Pa)
1	27.708	2.0360	51.715	0.068948	0.07030696	6894.8
0.036091	1	0.073483	1.8665	2.4884×10⁻³	2.537×10⁻³	248.84
0.491154	13.609	1	25.4 *	0.033864	0.034532	3386.4
0.0193368	0.53578	0.03937	1	0.0013332	0.0013595	133.32
14.5038	401.86	29.530	750.062	1	1.01972	10⁵*
14.223	394.1	28.959	735.559	0.98066	1	98,066.5*
1.45038×10⁻⁴	4.0186×10⁻³	2.953×10⁻⁴	0.00750	10⁻⁵	1.0192×10⁻⁵	1

1 atm (normal = 760 Torr) = 1.01325×10⁵ Pa
1 atm (technical = 1 kg$_f$/cm²) = 9.806650×10⁴ Pa*

Application Rate

lb/la-mi	lb/ft²	oz/yd²	kg/la-km	g/m²
1	1.58×10⁻⁵	2.273×10⁻³	0.282	7.706×10⁻²
63,360*	1	144*	1.79×10⁴	4882.5
440*	6.94×10⁻³	1	124	33.9
3.55	6.24×10⁻²	8.06×10⁻³	1	0.273
12.98	2.05×10⁻⁴	0.0295	3.66	1

1 lb/la-mi = 0.0771 g/m²; 1 g/m² = 13 lb/la-mi. (approx.)

Application Rate Equivalents of lb/la-mi

lb/la-mi	lb/ft²	oz/yd²	kg/la-km	g/m²
1	1.58×10⁻⁵	2.27×10⁻³	0.282	0.077
50	7.90×10⁻⁴	0.114	14.1	3.86
100	1.58×10⁻³	0.227	28.2	7.71
200	3.16×10⁻³	0.454	56.4	15.4
300	4.73×10⁻³	0.682	84.6	23.1
400	6.31×10⁻³	0.909	113	30.8
500	7.89×10⁻³	1.14	141	38.5

Application Rate Equivalents of g/m^2

g/m^2	lb/ft^2	lb/la-mi	kg/la-km	oz/yd^2
1	2.05×10^{-4}	13	3.66	0.0295
5	1.02×10^{-3}	65	18.3	0.148
10	2.05×10^{-3}	130	36.6	0.295
15	3.07×10^{-3}	195	54.9	0.442
20	4.10×10^{-3}	260	73.2	0.590
30	6.14×10^{-3}	389	110	0.885
40	8.19×10^{-3}	519	146	1.18
50	1.02×10^{-2}	649	183	1.47

Application Rate Equivalents of g/m^2
(Units Used for Airport Runways)

g/m^2	$kg/100{,}000\ ft^2$	$lb/100{,}000\ ft^2$	$kg/1000\ m^2$	$kg/10{,}000\ m^2$
10	93	205	10	100
15	140	310	15	150
20	185	410	20	200
25	235	520	25	250
30	280	620	30	300
40	370	815	40	400
60	560	1325	60	600

Liquid Application Area Coverage

$gal/1000\ ft^2$	$L/100\ m^2$
1	0.035168
28.435	1

Appendix A

Wind Unit Conversion Table

	Multiply by			
Convert from ⇒to	mi/h	m/s	km/h	kn
mi/h	—	0.44704	1.60934	1.15193
m/s	2.23693	—	3.6	1.96936
km/h	0.62137	0.27778	—	0.54704
kn	0.86811	0.50778	1.82800	—

Example: To convert 10 mi/h to m/s: 10×0.44704 = 4.47 m/s
To convert 10 m/s to mi/h: 10×2.23693 = 22.4 mi/h

Celsius to Fahrenheit Conversion

C	F	C	F	C	F	C	F
50	122.0	25	77.0	0	32.0	−25	−13.0
49	120.2	24	75.2	−1	30.2	−26	−14.8
48	118.4	23	73.4	−2	28.4	−27	−16.6
47	116.6	22	71.6	−3	26.6	−28	−18.4
46	114.8	21	69.8	−4	24.8	−29	−20.2
45	113.0	20	68.0	−5	23.0	−30	−22.0
44	111.2	19	66.2	−6	21.2	−31	−23.8
43	109.4	18	64.4	−7	19.4	−32	−25.6
42	107.6	17	62.6	−8	17.6	−33	−27.4
41	105.8	16	60.8	−9	15.8	−34	−29.2
40	104.0	15	59.0	−10	14.0	−35	−31.0
39	102.2	14	57.2	−11	12.2	−36	−32.8
38	100.4	13	55.4	−12	10.4	−37	−34.6
37	98.6	12	53.6	−13	8.6	−38	−36.4
36	96.8	11	51.8	−14	6.8	−39	−38.2
35	95.0	10	50.0	−15	5.0	−40	−40.0
34	93.2	9	48.2	−16	3.2	−41	−41.8
33	91.4	8	46.4	−17	1.4	−42	−43.6
32	89.6	7	44.6	−18	−0.4	−43	−45.4
31	87.8	6	42.8	−19	−2.2	−44	−47.2
30	86.0	5	41.0	−20	−4.0	−45	−49.0
29	84.2	4	39.2	−21	−5.8	−46	−50.8
28	82.4	3	37.4	−22	−7.6	−47	−52.6
27	80.6	2	35.6	−23	−9.4	−48	−54.4
26	78.8	1	33.8	−24	−11.2	−49	−56.2

Fahrenheit to Celsius Conversion

F	C	F	C	F	C	F	C
125	51.6	75	23.9	25	−3.9	−25	−31.6
124	51.1	74	23.3	24	−4.4	−26	−32.2
123	50.5	73	22.8	23	−5.0	−27	−32.7
122	50.0	72	22.2	22	−5.6	−28	−33.3
121	49.4	71	21.6	21	−6.1	−29	−33.9
120	48.8	70	21.1	20	−6.7	−30	−34.4
119	48.3	69	20.5	19	−7.2	−31	−35.0
118	47.7	68	20.0	18	−7.8	−32	−35.5
117	47.2	67	19.4	17	−8.3	−33	−36.1
116	46.6	66	18.9	16	−8.9	−34	−36.6
115	46.1	65	18.3	15	−9.4	−35	−37.2
114	45.5	64	17.8	14	−10.0	−36	−37.7
113	45.0	63	17.2	13	−10.5	−37	−38.3
112	44.4	62	16.7	12	−11.1	−38	−38.9
111	43.8	61	16.1	11	−11.7	−39	−39.4
110	43.3	60	15.5	10	−12.2	−40	−40.0
109	42.7	59	15.0	9	−12.8	−41	−40.5
108	42.2	58	14.4	8	−13.3	−42	−41.1
107	41.6	57	13.9	7	−13.9	−43	−41.6
106	41.1	56	13.3	6	−14.4	−44	−42.2
105	40.5	55	12.8	5	−15.0	−45	−42.7
104	40.0	54	12.2	4	−15.5	−46	−43.3
103	39.4	53	11.7	3	−16.1	−47	−43.8
102	38.9	52	11.1	2	−16.7	−48	−44.4
101	38.3	51	10.5	1	−17.2	−49	−45.0
100	37.7	50	10.0	0	−17.8	−50	−45.5
99	37.2	49	9.4	−1	−18.3	−51	−46.1
98	36.6	48	8.9	−2	−18.9	−52	−46.6
97	36.1	47	8.3	−3	−19.4	−53	−47.2
96	35.5	46	7.8	−4	−20.0	−54	−47.7
95	35.0	45	7.2	−5	−20.5	−55	−48.3
94	34.4	44	6.7	−6	−21.1	−56	−48.8
93	33.9	43	6.1	−7	−21.6	−57	−49.4
92	33.3	42	5.6	−8	−22.2	−58	−50.0
91	32.7	41	5.0	−9	−22.8	−59	−50.5
90	32.2	40	4.4	−10	−23.3	−60	−51.1
89	31.6	39	3.9	−11	−23.9	−61	−51.6
88	31.1	38	3.3	−12	−24.4	−62	−52.2
87	30.5	37	2.8	−13	−25.0	−63	−52.7
86	30.0	36	2.2	−14	−25.5	−64	−53.3
85	29.4	35	1.7	−15	−26.1	−65	−53.8
84	28.9	34	1.1	−16	−26.6	−66	−54.4
83	28.3	33	0.6	−17	−27.2	−67	−54.9
82	27.8	32	0.0	−18	−27.8	−68	−55.5
81	27.2	31	−0.6	−19	−28.3	−69	−56.1
80	26.6	30	−1.1	−20	−28.9	−70	−56.6
79	26.1	29	−1.7	−21	−29.4	−71	−57.2
78	25.5	28	−2.2	−22	−30.0	−72	−57.7
77	25.0	27	−2.8	−23	−30.5	−73	−58.3
76	24.4	26	−3.3	−24	−31.1	−74	−58.8

Appendix A

Wind-Chill Temperature Chart

Speed mi/h	Temperature, °F								
	50	40	30	20	10	0	−10	−20	−30
Calm	50	40	30	20	10	0	−10	−20	−30
5	48	37	27	16	6	−5	−15	−26	−36
10	40	28	16	4	−9	−21	−33	−46	−58
15	36	22	9	−5	−18	−36	−45	−58	−72
20	32	18	4	−10	−25	−39	−53	−67	−82
25	30	16	0	−15	−29	−44	−59	−74	−88
30	28	13	−2	−18	−33	−48	−63	−79	−94
35	27	11	−4	−20	−35	−49	−67	−82	−98
40	26	10	−6	−21	−37	−53	−69	−85	−100

The wind-chill temperature is a measure of relative discomfort due to co͟͟͟͟
It was developed by Siple and Passel (1941) and is based on physiologica͟͟͟
heat loss for various combinations of ambient temperature and wind spee͟͟͟
perature equals the actual air temperature when the wind speed is 4 mi/h ͟͟͟
speeds, the wind-chill temperature is lower than the air temperature and n͟͟͟
cold stress and discomfort associated with wind.

The effects of wind chill depend strongly on the amount of clothing and͟͟͟
as well as on age, health, and body characteristics. Wind-chill temperatu͟͟͟
indicate that there is a risk of frostbite or other injury to exposed flesh. Th͟͟͟
from being inadequately clothed also depends on the wind-chill temperature

Appendix B

Chemicals and Their Properties

The characteristics of the principal chemicals used for ice control, and also of some that have been used in the past but have fallen into disfavor because of environmental considerations, cost, or other reasons, are described in this appendix. The latter are included only for purposes of comparison and to provide a reference. This may prevent the unintentional use of chemicals that can damage the environment. Tables giving the most important characteristics of the chemicals described appear at the end of this appendix.

Calcium Chloride, $CaCl_2$

Though a pure grade of calcium chloride is produced, it is not used for ice control because its higher cost does not produce commensurate benefits. It is used as a food additive and as a reagent by analytical laboratories. The properties and characteristics of commercial grades may differ slightly from vendor to vendor. The differences will amount to only a few percentage points, so the figures given below (Tables B-3 and B-4) will be usable for the engineering calculations to be described.

Calcium chloride dissolves readily in water. One outstanding benefit of this chemical is its low eutectic temperature: It will continue to melt ice down to a temperature of −51°C (−60°F) at a concentration of about 30 wt%. At higher concentrations, several hydrates are formed (hydrates are stable forms of the molecule which contain one or more molecules of water bound to the chemical). The solubility (phase) diagram (Fig. B-1) shows these forms up to a temperature of 50°C (122°F). The shape of the

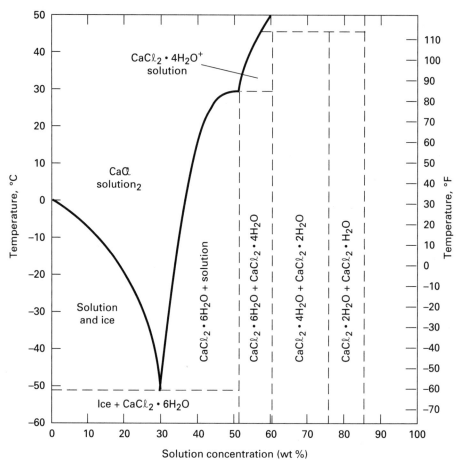

Figure B-1. Solubility diagram for calcium chloride.

solubility diagram is the result of the differing solubilities of the various hydrates. Only calcium chloride, in solution, will exist at all points above the solid line (called the liquidus, solution, or saturation curve); below that line, ice will be mixed with the solution up to a concentration of about 30 wt%, and beyond that concentration various hydrates will form. Note, however, that below the eutectic temperature of −51°C (−60°F), a solid mixture of ice and hexahydrate form of $CaCl_2$ results. The properties of the various hydrates are listed in Table B-1. Heats of solution are all exothermic (the hydrates liberate heat when dissolved in water) except for the hexahydrate, which absorbs heat from the solution.

Calcium chloride produced for ice control contains some impurities, which account for the slight differences in properties. Three forms are

Appendix B

Table B-1. Forms of Calcium Chloride

Formula	Description	% $CaCl_2$	Heat of solution* Btu/lb	Heat of solution* kJ/kg
$CaCl_2$	Anhydrous	100	317	737
$CaCl_2 \cdot H_2O$	Monohydrate	86.0	174	404
$CaCl_2 \cdot 2H_2O$	Dihydrate	75.5	131	304
$CaCl_2 \cdot 4H_2O$	Tetrahydrate	60.6	26	60
$CaCl_2 \cdot 6H_2O$	Hexahydrate	50.7	−31	−72

*Positive heat of solution releases heat, negative absorbs heat.

Table B-2. Forms of Commercial Calcium Chloride for Ice Control

Form	Composition	Concentration (% $CaCl_2$)	Bulk density kg/m^3	Bulk density lb/ft^3
Flake or pellet	Anhydrous	94–97	834–929	52–58
Flake or powder	Dihydrate	77–80	817–961	51–60
Liquid		30–45	—	—
		32 (example)	1320	82

available from several vendors. Table B-2 lists some useful characteristics compiled from manufacturers' literature (Allied Chemical, now General Chemical; Dow Chemical; Tetra Technologies). The values will vary slightly from vendor to vendor.

The natural state of calcium chloride is liquid. The solid form is obtained by heating the liquid chemical to drive off the water. This thermal energy is stored in the solid chemical and will be released when it returns to the liquid state. The heat of solution of calcium chloride used for ice control amounts to nearly 130 Btu/lb (300 kJ/kg) for the dihydrate and 315 Btu/lb (735 kJ/kg) for the anhydrous form. This is so high that if water is poured on the solid chemical when making a solution, it will heat so rapidly that it may boil and splatter onto you. For safety reasons, the chemical should always be added to water. Because of calcium chloride's aggressive pursuit of water, the solid form requires protection to prevent its return to the liquid state. Merely covering an open pile of calcium chloride may not be sufficient. Vaporproof containers are necessary for complete protection. This property is an advantage when calcium chloride is used for prewetting another ice control chemical such as salt because it will speed up the solution process.

Sources

Commercial $CaCl_2$ is obtained from two sources: by extraction from natural brines recovered from deep wells, principally in Michigan, and by a chemical manufacturing process, the Solvay process. In this process, calcium carbonate (limestone) reacts with sodium chloride to produce sodium carbonate (soda ash) and calcium chloride: $CaCO_3 + 2NaCl \rightarrow Na_2CO_3 + CaCl_2$.

Product Standard

The American Society for Testing and Materials (ASTM) has issued two standards for calcium chloride.

1. *D 98, Standard Specification for Calcium Chloride.* This specification establishes a classification system and indicates the chemical requirements that must be met, the ranges of sieve sizes that the various grades must pass, and packaging and shipping requirements.
2. *E 449, Standard Test Methods for Analysis of Calcium Chloride.* This document details the chemical testing procedures for determining the concentrations of calcium chloride, magnesium chloride, potassium chloride, sodium chloride, and calcium hydroxide in a sample of calcium chloride.

   Note: There is another ASTM specification for $CaCl_2$, D 345, *Standard Test Method for Sampling Calcium Chloride for Roads and Structural Applications,* which describes procedures for sampling and determining total chlorides ($CaCl_2$, $MgCl_2$, NaCl, KCl, etc.) rather than individual chlorides. It refers to E 449 for the complete analysis and to D 98 for chemical compliance.

ASTM publications can be obtained from American Society for Testing and Materials, 1916 Race Street, Philadelphia, PA 19103. Vendors may also be able to furnish them—ask.

Preparation of Liquid $CaCl_2$

Mixing directions should be followed carefully because of the high temperatures that will result when solid calcium chloride is mixed with water. Note carefully the Caution flag in the mixing directions below. The solid dissolves readily in cold water, with little agitation

Appendix B

required. To obtain a specific concentration of solution if the volume of the container is known, follow the procedure of Method 1. If the volume of the mixing container is not known, use Method 2. The procedures for English and for metric units are given separately. This information is adapted from the *Calcium Chloride Handbook* (Form No. 173-01534-396 AMS) from The Dow Chemical Company. Similar information is available from all commercial $CaCl_2$ sources.

English Units

Method 1: If the Container Capacity Is Known

1. From column 1 or column 3 of Table B-3, determine the quantity of solid calcium chloride required to make one gallon of solution. This value multiplied by the volume of the container gives the total weight of calcium chloride required.
2. Fill the container approximately two-thirds full of cold water, than add the required calcium chloride *gradually* while stirring gently. You can use a paddle stirred by hand, a mechanical agitator, or an air bubbler.
3. After the calcium chloride has completely dissolved, add water to bring the level in the container to the working volume, then agitate slowly until a uniform mixture is obtained.

Example. You have a container with a known 300-gal capacity to its fill line, and you want to make that amount of 20 percent solution.

Table B-3. Calcium Chloride Mixing Proportions

Percent $CaCl_2$ actual	Pounds $CaCl_2$, 78% flake		Pounds $CaCl_2$, 95% pellet		Crystallization starts (°F)	Weight of 1 gal solution (lb)
	Col. 1: per gallon solution	Col. 2: per gallon water	Col. 3: per gallon solution	Col. 4: per gallon water		
10	1.16	1.22	0.95	0.97	22.3	9.06
15	1.82	1.99	1.50	1.55	13.5	9.47
20	2.53	2.87	2.10	2.22	−4.0	9.88
25	3.31	3.93	2.72	2.92	−21.0	10.3
29.87*	4.10	5.18	3.39	3.81	−67	10.8
30	4.16	5.23	3.42	3.84	−50.8	10.8

*This is the eutectic point, i.e., the concentration which results in the lowest temperature (−67°F) at which a solution can exist while remaining completely liquid.

Put about 200 gal of water (two-thirds of the container's capacity) in the container and add 759 lb of flake (77–80 percent) $CaCl_2$ [300 gal×2.53 lb $CaCl_2$/gal solution (read from column 1)] or 630 lb of pellet (94–97 percent) $CaCl_2$ [300 gal×2.1 lb $CaCl_2$/gal solution (read from column 3)]. When the $CaCl_2$ is dissolved, add water to the 300-gal mark and mix completely.

Method 2: If the Container Volume Is Not Known

1. Put a measured volume of water in the container. Do not exceed two-thirds of the container capacity, since adding the $CaCl_2$ will increase the solution volume by an unknown amount.
2. Add the number of pounds of calcium chloride called for in column 2 or column 4 of Table B-3 for each gallon of water used. Slowly add the calcium chloride to the water while agitating.
3. When the chemical is completely dissolved, the solution will have the desired concentration.

Example. You don't know the capacity of your container, but you find you can put in 200 gal of water and the container is no more than two-thirds full. You want to make a 30 percent solution. Add 1046 lb of flake $CaCl_2$ to the 200 gal [200 gal×5.23 lb/gal water (read from column 1)]. When the chemical is completely dissolved, the volume will have increased above the mark to which you first filled it, and the concentration will be 30 percent.

At 30% concentration, crystals of the hexahydrate, $CaCl_2 \cdot 6H_2O$, will form. Of course, as the solution becomes diluted, as would be the case when it is used to melt snow or ice on the road, the concentration decreases, and the freezing point rises. As an example, to avoid freezing of the solution at a temperature of 22°F, the concentration must not drop below 10 percent.

Caution. Calcium chloride is exothermic: It gives off heat when dissolved in water. This heat of solution causes the brine to expand and occupy more space than it will after it cools. That's why the container in Method 2 is filled to no more than two-thirds capacity. For example, additional tank capacity of approximately 23 gal for every 1000 gal of 20 percent solution is required. This will increase to 26 gal for every 1000 gal of a 34 percent solution. In the example given above for Method 2, the 200 gal will expand to about 205 gal.

Always add the calcium chloride to the water. If you put the calcium chloride in the container first and then add water, the chemical may form a solid mass which is difficult to dissolve completely.

Appendix B

Calculation Method

G = Gallons of water required to make a solution of a desired concentration

$$G = \left(\frac{\text{lb dry CaCl}_2 \times \%\text{CaCl}_2}{\text{desired \% solution}} - \text{lb dry CaCl}_2 \right) \div 8.34$$

Example. To make a 20 percent solution from 1000 lb of flake $CaCl_2$ (this is typically 78 percent concentration),

$$G = \left(\frac{1000 \times 78}{20} - 1000 \right) \div 8.34 = 348 \text{ gal water}$$

Increasing the Concentration of a Solution. If the concentration of a solution you have prepared is not high enough, more $CaCl_2$ can be added to bring it to the desired concentration. Use this equation to calculate the weight of $CaCl_2$ to add.

$$\frac{\%\text{ concentration desired} - \%\text{ concentration weak solution}}{\%\text{ CaCl}_2 \text{ in solid form} - \%\text{ concentration desired}}$$

$$\times \text{ weight of weak solution (lb/gal)} = \text{lb solid CaCl}_2 \text{ to add}$$

Example. To increase the concentration of a 20 percent solution to 30 percent,

$$\frac{(30 - 20) \times 9.88}{78 - 30} = \frac{2.06 \text{ lb of 78\% CaCl}_2}{\text{to add for each gallon of weak solution}}$$

Decreasing the Concentration of a Solution. If the concentration of a solution you have prepared is too high, dilute the strong solution by adding more water. Use this calculation to determine the amount of water to add.

$$\frac{\%\text{ concentration of strong solution} - \%\text{ concentration of weak solution}}{\%\text{ concentration of weak solution}}$$

$$\times \frac{\text{weight of strong solution}}{8.34} = \text{gal water to add}$$

Example. To dilute a 30 percent solution to 20 percent,

$$\left(\frac{30 - 20}{20} \right) \times \left(\frac{10.8}{8.34} \right) = \frac{0.65 \text{ gal of water to add}}{\text{for each gal of 30 percent solution}}$$

Metric Units

Method 1: If the Container Capacity Is Known

1. From column 1 or column 3 of Table B-4, determine the quantity of solid calcium chloride required to make one liter of solution. This value multiplied by the volume of the container gives you the total weight of calcium chloride required.
2. Fill the container approximately two-thirds full of cold water, then add the required calcium chloride *gradually* while stirring gently. You can use a paddle stirred by hand, a mechanical agitator, or an air bubbler.
3. After the calcium chloride has completely dissolved, add water to bring the level in the container to the working volume, then agitate slowly until a uniform mixture is obtained.

Example. You have a container with a known 1000-L capacity to its fill line, and you want to make that amount of 20 percent solution. Put about 650 L of water (two-thirds of the container's capacity) in the container and add 304 kg of flake (77–80 percent) $CaCl_2$ [1000 L×0.304 kg $CaCl_2$/L solution (read from column 1)] or 250 kg of pellet (94–97 percent) $CaCl_2$ [1000 L×0.250 kg $CaCl_2$/L solution (read from column 3)]. When the $CaCl_2$ is dissolved, add water to the 1000-L mark and mix completely.

Table B-4. Calcium Chloride Mixing Proportions

Percent $CaCl_2$ actual	Kilograms $CaCl_2$, 78% flake		Kilograms $CaCl_2$, 95% pellet		Crystallization starts (°C)	Weight of 1 L solution (kg)
	Col. 1: per liter solution	Col. 2: per liter water	Col. 3: per liter solution	Col. 4: per liter water		
10	0.139	0.147	0.114	0.117	−5.4	1.087
15	0.218	0.237	0.179	0.187	−10.3	1.136
20	0.304	0.344	0.250	0.266	−18.0	1.185
25	0.396	0.470	0.325	0.356	−29.4	1.236
29.87*	0.496	0.618	0.407	0.456	−55.0	1.293
30	0.498	0.623	0.409	0.460	−46.0	1.295

*This is the eutectic point, i.e., the concentration which results in the lowest temperature (−55°C) at which a solution can exist while remaining completely liquid.

Appendix B

Method 2: If the Container Volume Is Not Known

1. Put a measured volume of water in the container. Do not exceed two-thirds of the container capacity, since adding the $CaCl_2$ will increase the solution volume by an unknown amount.
2. Add the number of kilograms of calcium chloride called for in column 2 or column 4 of Table B-4 for each liter of water. Slowly add the calcium chloride to the water while agitating.
3. When the chemical is completely dissolved, the solution will have the desired concentration.

Example. You don't know the capacity of your container, but you find you can put in 500 L of water and the container is no more than two-thirds full. You want to make a 30 percent solution. Add 311.5 kg of flake $CaCl_2$ to the 500 L [500 L×0.623 kg/L water (read from column 2)] or add 230 kg of pellet $CaCl_2$ to the 500 L water [500 L×0.460 kg/L (read from column 4)]. When the chemical is completely dissolved, the volume will have increased above the mark to which you first filled it, and the concentration will be 30 percent.

At 30 percent concentration, crystals of the hexahydrate, $CaCl_2 \cdot 6H_2O$, will form. Of course, as the solution becomes diluted, as would be the case when it is used to melt snow or ice on the road, the concentration decreases, and the freezing point rises. As an example, to avoid freezing of the solution at a temperature of −5°C, the concentration must not drop below 10 percent.

Increasing the Concentration of a Solution. If the concentration of a solution you have prepared is not high enough, more $CaCl_2$ can be added to bring it to the desired concentration. Use this equation to calculate the weight of $CaCl_2$ to add.

$$\frac{\% \text{ concentration desired} - \% \text{ concentration weak solution}}{\% \ CaCl_2 \text{ in solid form} - \% \text{ concentration desired}}$$

$$\times \text{ weight of weak solution (kg/L)} = \text{kg solid } CaCl_2 \text{ to add}$$

Example. To increase the concentration of a 20 percent solution to 30 percent,

$$\frac{(30 - 20) \times 1.185}{78 - 30} = \begin{array}{l} 0.25 \text{ kg of 78\% } CaCl_2 \\ \text{to add for each liter of weak solution} \end{array}$$

Decreasing the Concentration of a Solution. If the concentration of a solution you have prepared is too high, dilute the strong solution by adding more water. Use this calculation to determine the amount of water to add.

$$\left(\frac{\%\text{ concentration of strong solution} - \%\text{ concentration of weak solution}}{\%\text{ concentration of weak solution}}\right)$$
$$\times \text{ weight of strong solution} = \text{liters water to add}$$

Example. To dilute a 30 percent solution to 20 percent,

$$\left(\frac{30-20}{20}\right) \times 1.295$$
$$= 0.65 \text{ liter of water to add for each liter of 30 percent solution}$$

Sodium Chloride, NaCl

Salt is the mineral halite (from the Greek *hals*, salt) that occurs as isometric cubic crystals. In its pure form it is one of the most stable of common chemicals. That means that it will not undergo change in form or composition under normal conditions of temperature and pressure. As an indication of this stability, many rock salt deposits throughout the world are millions of years old, deposited through countless centuries in ancient seas. As long as they are kept under covered storage, salt stockpiles will not deteriorate but will remain unchanged indefinitely. There is no change in the melting effectiveness. Only when salt is dissolved in water will its melting effectiveness change: As a salt solution becomes more dilute, its freezing point will rise. This means that it will melt less snow or ice, and it will freeze at a higher temperature. If stockpiled salt is exposed to rain or snow, it will dissolve, but if the solution is collected and the water evaporated, the salt will return to its original state.

Sources

Three methods are used for the production of salt: *Rock salt* is mined by conventional hard rock mining equipment and techniques, *solar salt* is produced by the evaporation of seawater, and *solution salt* is obtained by injecting water into deep underground deposits and pumping out the dissolved salt. Most salt used for highway applications in the United States is rock salt, although some solar salt produced in California and Utah and some imported into the eastern United States is used for ice control. Naturally occurring rock salt usually contains between 1 and 4 percent impurities, mostly insoluble anhydrite ($CaSO_4$), gypsum ($CaSO_4 \cdot 2H_2O$), shale, dolomite, calcite

Appendix B

($CaCO_3$), pyrite, iron oxides, and quartz (Kaufman 1960).

Material Standards

The ASTM designation for salt used for highway ice control is D 632, *Standard Specification for Sodium Chloride*. In addition, E 534, *Standard Test Method for Chemical Analysis of Sodium Chloride*, details the methods for sample preparation and for determining the amount of insolubles and the calcium, magnesium, and sulfate contents.

Gradations

Two gradations of salt, fine and coarse, are designated by the ASTM standard. Their sizes are given in Table B-5.

Instructions for Mixing Selected Concentrations of Salt Solutions

The weight of salt to add to water to obtain a concentration (in weight percent) which has a desired freezing point depression is listed in Table B-6 for several concentrations. This chart is adequate for most uses, but if intermediate concentrations or specific gravities are required to aid in bulk mixing, use Table B-7. The solubility diagram (phase diagram), Fig. B-2, is also useful for selecting a concentration of salt for obtaining a specified freezing point (this figure repeats Fig. 3-3 in Chapter 3, where guidance on how to interpret the chart can be found).

Table B-5. Gradation of Salt Specified by ASTM D632.

Sieve size	Weight % passing	
	Grade 1*	Grade 2
3/4 in. (19.0 mm)	—	100
1/2 in. (12.5 mm)	100	—
3/8 in. (9.5 mm)	95–100	—
No. 4 (4.75 mm)	20–90	20–100
No. 8 (2.36 mm)	10–60	10–60
No. 30 (600 μm)	0–15	0–15

*Grade 1 is most commonly used in the United States.

Table B-6. Proportions for Preparing Sodium Chloride Solutions

Percent NaCl actual	Pounds NaCl Col. 1: per gallon solution	Pounds NaCl Col. 2: per gallon water	Freezing begins (°F)	Weight of 1 gal solution (lb)
10	0.895	0.834	+20.2	8.95
15	1.39	1.25	+12.3	9.28
20	1.92	1.67	−0.4	9.60
23.3*	2.29	1.94	−6.02	9.76
25	2.488	2.08	+16.1	10.3

*This is the eutectic composition and temperature.

Table B-7. Salt Concentration and Corresponding Specific Gravity (Measured by a Hydrometer) at Temperatures Shown

Percent salt	Specific gravity at 59°F (15°C)	Percent of saturation	Weight of salt (lb/gal)	Specific gravity at 32°F (0°C)
0	1.000	0	0	
5	1.035	20	0.432	1.038
6	1.043	24	0.523	1.046
7	1.050	28	0.613	1.054
8	1.057	32	0.706	1.061
9	1.065	36	0.800	1.069
10	1.072	40	0.895	1.077
11	1.080	44	0.992	1.085
12	1.087	48	1.000	1.092
13	1.095	52	1.100	1.100
14	1.103	56	1.291	1.108
15	1.111	60	1.392	1.116
16	1.118	63	1.493	1.124
17	1.126	67	1.598	1.323
18	1.134	71	1.705	1.140
19	1.142	75	1.813	1.149
20	1.150	79	1.920	1.157
21	1.158	83	2.031	1.165
22	1.166	87	2.143	1.173
23	1.175	91	2.256	1.182
24	1.183	95	2.371	1.190
25	1.191	99	2.448	
25.2	1.200	100		

Weight of commercial salt required = (weight of pure NaCl from table)÷(purity in percent).

Appendix B

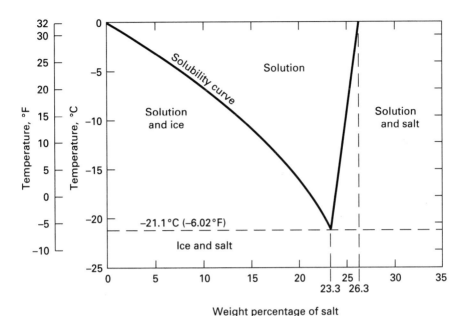

Figure B-2. Solubility diagram for sodium chloride.

Additives

The critical humidity for salt is about 73 percent. This means that salt will absorb moisture from the air and go completely into solution at any relative humidity above this value. Below that level of humidity, a film of saturated brine is present on salt crystals that are stored in contact with air. This may cause caking, which can result in formation of lumps and otherwise interfere with free flowing in loading and spreading equipment. A chemical is frequently added to salt to make it noncaking or to keep it free-flowing.* One of two chemicals is used commercially: Prussian blue [ferric ferrocyanide, $Fe_4(Fe(CN)_6)_3 \cdot xH_2O$] or yellow prussiate of soda [YPS; sodium ferrocyanide decahydrate, $Na_4Fe(CN)_6 \cdot 10H_2O$]. Prussian blue is generally added at the rate of 70 to 165 ppm (0.14 to 0.33 lb/ton or 70 to 165 g/†, where 1† = 1000 kg), and YPS at the rate of 50 to 100 ppm (0.1 to 0.2 lb/ton or 50 to 100 g/†). Both of these chemicals are permitted as additives for human uses, Prussian blue for coloring drugs used for external application (21CFR73.1299)† and YPS as an anticaking additive for table salt (21CFR172.490)†) (Kuhajek and Fiedelman 1976).

---

*"Noncaking" and "free-flowing" have different meanings. Free-flowing means that the salt flows down chutes without interference such as arching or sticking, whereas noncaking means that the salt crystals do not cement together in lumps in storage and does not imply free-flowing (Kaufmann 1978).

†These are the Code of Federal Regulations citations (U.S. Government Printing Office).

Environmental Effects of Salt

Salt can injure roadside vegetation through two mechanisms: increasing the soil salt concentration, which can result in salt absorption through roots that places them under osmotic stress, and accumulation of salt on leaves, needles, and branches as a result of splash and spray (TRB 1991). Though much of the surface drainage from highways may flow into streams during the winter, salt ions may still infiltrate frozen soils, and often soils are protected by heavy snow cover and do not freeze (Hanes et al. 1976).

The salt molecule NaCl readily dissolves in water and breaks up into its constituents, the sodium ion Na^+ and the chlorine ion Cl^-. When salty water leaches into the ground, the sodium ion can displace the potassium, calcium, and magnesium in clays and adversely affect the soil structure. Experience in Maine has demonstrated the effectiveness of applying up to 14 tons/acre of gypsum (calcium sulfate, $Ca_2SO_4 \cdot 2H_2O$) for improving the soil infiltration and moisture retention of soil whose structure has been seriously altered by heavy salt runoff. It was found that limestone, a mineral consisting principally of calcium carbonate, did not reduce sodium in the soil. This occurs because when sodium is replaced by calcium in the presence of carbonate ions, an insoluble sodium carbonate forms. However, when sodium is replaced by calcium in the presence of sulfate ions, as occurs when gypsum is applied, the result is the formation of a highly soluble sodium sulfate which leaches through the soil when it rains (Hutchinson 1972).

Selection of Salt-Tolerant Plants

Table B-8 lists trees, both deciduous and coniferous, and shrubs classified by their relative tolerance to salt under winter conditions (Welch 1977, Sucoff 1975). "Spray" indicates that the salt is deposited on the leaves or needles by splash or spray through air; "soil" indicates that the salt from the runoff is absorbed through the roots.

Magnesium Chloride, $MgCl_2$

Magnesium chloride has many characteristics that are similar to those of calcium chloride (see Table B-9): It absorbs moisture from the air (hygroscopicity), it gives off heat when dissolved in water (exothermicity), and it has nearly the same ice-melting capacity as $CaCl_2$. Large quantities of $MgCl_2$ are used for the production of magnesium metal and in the preparation of oxychloride cements. Other major commer-

Table B-8. Trees (Deciduous and Coniferous) and Shrubs Classified by their Relative Tolerance to Salt under Winter Conditions

LATIN NAME	COMMON NAME	Spray	Soil Na	Soil Cl	General	COMMENTS
Abies balsamea	Balsam Fir		MT		VS	
Acacia			MT		VS	
Acanthopanax sieboldianus	Five Leaf Aralia				VS	KEY:
Acer species	Maple	VT, MT			VS, S	VS = Very Sensitive
Acer campestre	Hedge Maple				MT, T	S = Sensitive
Acer carpenifolia	Hornbeam Maple	S				MT = Moderately Tolerant
Acerdavidii	David Maple	S				T = Tolerant
Acer ginnala	Aunr Maple	VS				VT = Very Tolerant
Acer grosseri v. hersii	Grassers Maple	S				
Acer Mono	Mono Maple	T				
Acer Monspessulanum	Montpelier Maple	S				
Acer negundo	Box Elder	MT, VS	MT		MT, MT	
Acer palmatum	Japanese Maple	VS				
Acer platanoides	Norway Maple	VT, T, T	MT		MT, T	
Acer pseudoplatanus	Sycamore Maple	T	VS		VS, S, VS	
Acer rubrum	Red Maple	MT	VS		VS	
Acer saccharinum	Silver Maple	MT, T				
Acer saccharum	Sugar Maple	T	MT		VS	
Acer sieboldianum	Siebold Maple	MT				
Acer spicatum	Mountain Maple	VS				
Acer tataricom	Tatarian Maple	VS				
Acer velutinum	Velvet Maple	VT			VT, S	
Aesculus hippocastanum	Common or White Horsechestnut	VT			MT, VT	
Ailanthus attissima	Tree of Heaven	VT			MT	
Alnus spe	Alder					
Alnus glutinosa	European Alder	T, MT			MT	
Alnus hirsuta	Manchurian Alder	T				
Alnus incana	Speckled Alder	MT	VS		VS, S, vS	
Alnus rugosa	Speckled Alder	MT			S	
Amelanchier asiatica	Asian Serviceberry	VS				

Table B-8 (Cont).

LATIN NAME	COMMON NAME	Spray	Soil Na	Soil Cl	General	COMMENTS
Amelanchier canadensis	Shodblow Serviceberry	VS				
Amelanchier grandiflora	Apple Serviceberry	VS				
Amelanchier laevis	Allegany Serviceberry	T				
A. ovalis	Garden Serviceberry	S				
A. spicata	Dwarf Serviceberry					
Arctostaphylos uva-ursi	Bearberry	T			S	Prefers sandy soils. Can be difficult to transplant. Vigorous spreader. No special attention required. No insect or disease problems.
Aronia arbutifolia	Red Chokecherry				MT	
Aronia melanocarpa	Black Chokecherry				MT	
Artemisia abrotanum 'Nara'	Dwarf Southernwood				T	Requires reasonably fertile soil tolerates alkaline soils. Require little care. Some dying out due to disease.
Atriplex species	Salt Bush				T	
Atriplex patola (hastata)	Eastern Baccharus Groundsel-				VT	
Baccharis halimifolia	Bush				VT	Needs moist, fairly saline soils
Berberis species	Barberry	S			VS	
Berberis aggregate	Salmon Barberry	T				
Berberis beanlana	Beans Barberry					
Berberis bretschneideri	Purpleberry Barberry	VS				
Berberis dictyophylla	Chokleaf Barberry	VS				
Berberis dielsiana "Compacta"	Dwarfdiels Barberry	VS				
Berberis fendleri	Colorado Barberry	MT				
Berberir gagnepaini	Black Barberry	VS				
Berberis giraldii	Girald Barberry	VS				
Berberis hookeri	Hookers Barberry	VS				
Berberis julianae	Wintergreen Barberry	VS				
Berberis koreana	Korean Barberry	MT				
Berberis oblonga	Bigflower Barberry	MT				

Table B-8 (Cont.).

LATIN NAME	COMMON NAME	Spray	Soil Na	Cl	General	COMMENTS
Berberis thibetica	Thibetan Barberry	VS				Available. Woody ornamental.
Berberis thunbergii	Japanese Barberry	VS				Needs regular cultivation.
Berberis t. atropurpurea	Redleaf Ji Barberry	T				
Berberis tischleri	Tischler Barberry	VS				
Berberis vulgaris	Common Barberry	S				
Betula species	Birch	VS	MT		MT, VS	
Betula alba	White Birch				S	
Betula alleghaniensis	Yellow Birch				T	
Betula davoriea	Dahuvian Birch				VS, VS	
Betula ermannii	Ermans Birch	MT				
Betula humilis		S				
Betula lenta	Black Birch	MT			T	
Betula papyrifera	Paper or Canoe Birch	MT	MT		S, T	
Betula pendula	European Birch	MT			T	
Betula populifolia	Gray Birch					
Betula pubeicers		VS				
Betula schmidtii	Schmidts Birch	VS				
Buxus microphylla koreana	Koreana Boxwood				VS	Will not withstand roadside
Buxus sempervirens	Common Box		VS		VS, VS, S	environments
Callistemon lanceblatus	Bottle Brush		T			
Caragana arborescens	Siberian pea-tree	VT	VT, T		T	
Caragana frutex	Russian Peashrub				T	Large, woody shrub. Available
Carpinus betulus	European Hornbean	VS, S	VS		VS, VS	
Carpinus betulus quercifolia	Oakleaf Hornbean	S				
Carpinus caroliniana	American Hornbean				S, VS	
Carya species	Hickory				S	
Carya ovata	Shagbark Hickory	T			VS	
Carya pecan	Pecan				S	
Castanea dentata	American Chestnut	S				
Catalpa speciosa	Catalpa	MT				

Table B-8 (Cont.).

LATIN NAME	COMMON NAME	Spray	Soil Na	Soil Cl	General	COMMENTS
Celastrus orbiculata	Oriental Bittersweet	VS				Needs good soil on banks
Celtis caucasica	Caucasian Hackberry					
Cercis caradensis	Eastern redbud	S			MT	
Chaenomeles lagenaria	Flowering Quince					
Chamaecyparis pisifera	Swara False/cypress	S			MT	
Clematis orientalis					S	Not hardy enough for roadside use
Clematis panicubta					S	Not hardy enough for roadside use
Clematis tangutica					S	Not hardy enough for roadside use
Clematis texensis					S	Not hardy enough for roadside use
Clematis virginiana	Virginsbower					Ground cover, best in full sun. Available
Clethra alnifolia	Summersweet Clethra				S	
Colutea arborescens	Sweet Pepper Bush					
Colutea persica	Bladder-senna	T				Not Hardy
Comptonia peregrina	Persian Bladder-senna Sweetfern				T	Excellant, native, ground cover. Acid soils and full sun. Available
Condalia spathula	Knifeleaf Condalia	S			VT	
Cornus alba	Tatarian (Squawbush) Dogwood	VS				
Cornus alba 'Kesselringii'	Purpletwig Dogwood	VS				
Cornus alba 'Siberica'	Siberian Dogwood	VS			T	
Cornus alba 'Spaethii'	Yellowedge Dogwood					
Cornus amomum	Silky Dogwood	VS				
Cornus florida	Flowering Dogwood	VS			T	
Cornus mas	Cornelian Cherry	S			MT	
Cornus sanguinea	Gray Dogwood	VS,S			VS	
Cornus stelonifera	Redosier Dogwood	VS,S			MT	Spreading shrub or hedge. Adapted to all but dry climates
Corylus species	Filbert		VS		S	
Corylus americana	American Filbert	VS				

Table B-8 (Cont.).

LATIN NAME	COMMON NAME	Spray	Soil Na	Cl	General	COMMENTS
Corylus arellana	Common Filbert	VS,VS,S			VS,VS,VS	
Corylus colurnaides	Colurnoid Filbert	VS				
Corylus cornuta	Beaked Filbert	VS				
Corylus sieboldiana	Japanese Filbert	VS				
Cotoneaster diraricata	Spreading Cotoneaster	VS			S	
Cotoneaster integerrimus	European Cotoneaster	S	T		S,T	
Crataegus species	Hawthorn					
Crataegus coccinea (intricata)	Thicket Hawthorn	S			T	
Crataegus coccinoides	Kansas Hawthorn	VS				
Crataegus crusgalli	Cockspur Hawthorn	VS			T	
Crataegus grignonensis		VS				
Crataegus intrecata	Thicket Hawthorn	S			T	
Crataegus lavallei	Lavalle Hawthorn				MT,VS	
Crataegus mollis	Downy Hawthorn	S,S,VS			VS	
Crataegus monogyra	Singleseed Hawthorn	S,S,VS			T	
Crataegus oxyacantha	English Hawthorn					
Crataegus phaenopyrum	Washington Hawthorn					
Crataegus prunifolia	Plumleaf Hawthorn	VS				
Crataegus punctata	Dotted Hawthorn	S				
Crataegus sanguinea	Redhaw Hawthorn	S				
Crataegus sorbifolia	Sorbus Hawthorn	VS				
Cryptomeria japonica	Japanese Cryptomeria				MT	
Cydonia oblonga	Quince					
Cytisus purpurea procumbens	Prostrate Purple Broom	MT			S	Recommended only where it can be maintained.
Cytisus species	Broom				NS	Recommended on dry banks-all soils
Dievilla lonicera	Dwarf Bush Honeysuckle				VT	Hardy in full sun or shade
Diervilla rivularis	Georgia Bush Honeysuckle				VT	More vigorous than Dwarf Bush Honeysuckle. Available

Table B-8 (Cont.).

LATIN NAME	COMMON NAME	Spray	Soil Na	Cl	General	COMMENTS
Dervilla sessilifolia	Southern Bush Honeysuckle				VT	More vigorous than Dwarf Bush Honeysuckle. Available
Elaeagnus angustifolia	Russian Olive	T,T,T	T,T		MT,VT	
Elaeagnus commutata	Silverberry	VT			VT,VT	
Elaeagnus ebbingei		VS				
Elaeagnus multiflora	Cherry elaeagnus	MT				
Elaeagnus pungens reflexa		VS				
Elaeagnus pungens simoni		VS				
Elaeagnus umbellata	Autumn Elaeagnus	VS			T	
Eucalyptus globulus	Tasmanian Blue Eucalyptus				MT	
Eucalyptus robusta	Beakpod Eucalyptus				MT	
Euonymus alatus	Winged Euonymus (Burningbush)	T	VS,VS		T,VS	
Euonymus europaeus	European Euonymus (Spindletree)	S			T	
Euonymus fortunei vegetus	Bigleaf Wintercreeper				S	
Euonymus latifolius	Broadleaf Euonymus	S				
Euonymus nanus	Dwarf Euonymus	VS				
Euonymus verrucosus	Wartybark Euonymus	VS				
Eupatorium coelertinum	Mistflower Eupatorium	MT,VS				
Fagus grandifolia	American Beech	VS	MT		VS,S	
Fagus orientalis	Oriental Beech					
Fagus silvatica	European Beech	VS,S	VS		VS,MT	
					VS,S	
Fagus silvatica laciniata	Cutleaf European Beech	VS				
Fagus silvatica quericifolia	Oakleaf European Beech	VS				
Forsythia 'Arnold Dwarf'	Arnold Dwarf Forsythia				S	Needs fertile soils
Forsythia 'Bronxensis'	Bronx Forsythia				S	Not Hardy
Forsythia Hybrid	Hybrid Forsythia				S	Not Hardy
Forsythia intermedia	Border Forsythia	MT				
Forsythia intermedia 'Spectabilis'	Spring Glory Forsythia					
Fraxinus americana	White Ash	T	MT			
Fraxinus angustifolia	Narrowleaf Ash	MT				

Table B-8 (Cont.).

LATIN NAME	COMMON NAME	Spray	Na (Soil)	Cl (Soil)	General	COMMENTS
Fraxinus excelsior	European Ash	T,T			MT	
Fraxinus lidotricha		S				
Fraxinus ornus	Flowering Ash	VT				
Fraxinus pennsylvanica lanceolata	Green Ash		S,MT		MT,TT	
Gleditsia japonica	Japanese Honeylocust		MT		VT	
Halimodendron halodendron	Salttree (Siberian)	T	T,VT		T	
					VT	
Hemerocallis fulva	Tawny Daylily				T	Hardy perennial lily-like plant. 3 feet high easily grown. Readily available
Hibiscus syriacur	Shrubalthea	MT,T			T,T	
Hippohpae rhamnoides	Common Seabuck Thorn	MT			T,T	
Hydrangea macrophylla	Bigleaf Hydrangea				S	
Ilex aquifolium	English Holly	VS			MT	
Ilex crenatoa	Japanese Holly				S	
Ilex glabra	Inkberry				S	
Ilex opaca	American Holly				MT	
Ilex verticillata	Winterberry Black Alder		VS		MT,S	
Juglans species	Walnut				S	
Juglans nigra	Black Walnut	T	VS		VS	
Juglans regia	English or Persian Walnut	T			MT	
Juniperus species	Juniper	MT				
Juniperus chinensis pfitceriana	Pfitzer Juniper		T		T	Widely available, hardy.
Juniperus communis depressa	Prostrate Juniper				MT	Spreader, Needs further evaluation. Not too available
Juniperus conferta	Shore Juniper				VT	Excellent salt resistance; in some locations, not hardy
Juniperus horizontalis	Spreading (Creeping) Juniper				T	Prefers sandy dry soil. Requires little attention
Juniperus horizi 'Alpina'	Alpine Juniper		MT,MT		T	
Juniperus horizontalis	Andorra Creeping Juniper					

Table B-8 (Cont.).

LATIN NAME	COMMON NAME	Spray	Soil Na	Cl General	COMMENTS
Juniperus sabina 'Arcadia'	Arcadia Juniper				Arcadia & Skandia
Juniperus sabina 'Skandia'	Skandia Juniper	MT		MT	
Juniperus virginiana	Eastern Red Cedar		MT	MT, TT, T	
Kalmia latitolia	Mountain Laurel			MT	
Kolkwitzia amabilis	Beauty Bush	S			
Laburnum anagyroides	Golden Chain	VS			
Lantana camera	Lantana		MT		
Larix species	Larch			VS, VS	
Larix decidua	European Larch	T			
Larix larieina	Tamarock	T			
Larix leptolepis	Japanese Larch	T			
Lespedeza japonica intermedia	Japanese Lespedeza				
Ligustrum species	Privet	MT		MT, T	
Ligustrum amvrense	Amur Privet			MT	
Ligustrum ibotium	Ibolium Privet			MT	
Ligustrum obtusifolium regelionum	Regal Privet			T	
Ligustrum texanum	Texas Privet	MT, VS, S	MT		
Ligustrum vulgare	European Privet		VS	S, VS	
Liriodendron tulipitera	Tulip Tree	T			
Lonicera species	Honeysuckle			MT	
Lonicera amoena 'Alba'	White gotha Honeysuckle	S			
Lonicera caprifollum	Sweet Honeysuckle	S			
Lonicera coerulea	Sweetberry Honeysuckle	S			
Lonicera fragvantissima	Winter Honeysuckle			T	
Lonicera involucrata	Bearberry, Honeysuckle	VS			Not fully hardy
Lonicera japonica halliana	Halls Japanese Honeysuckle		MT, MT	T, MT, VT, T	
Lonicera ledebourii	Ledebour Honeysuckle	S			
Lonicera maackii	Amur Honeysuckle	S		T	
Lonicera morrowii	Morrow Honeysuckle	S			
Lonicera nigra		MT			

Table B-8 (Cont.).

LATIN NAME	COMMON NAME	Spray	Soil Na	Cl	General	COMMENTS
Lonicera periclymenum	Woodbine Honeysuckle	T				Not hardy
Lonicera purpusii	Purpus Honeysuckle	VS				
Lonicera syringantha	Lilac Honeysuckle	VS				
Lonicera tatarica	Tatarian Honeysuckle	VS	T		VT	
Lonicera xylosteum	European Fly Honeysuckl	MT, T			T	
Lycium chinensis	Chinese Waltberry				MT	Does not compete well with grasses and weeds
Lycium halinifolium (vulgare)	Matrimony Vine				VT, MT	
Malus species	Apple	S			VT	
Malus species	Crabapple	MT	MT		S	
Malus baccata	Siberian Crabapple				S	
Malus silvestris	Apple	VS	T			
Mespilus germanica	Medlar	VS				
Metasequoia glyptostroboides	Dawn Redwood				VT, T	
Morus species	Mulberry		T		VT	
Morus alba	White Mulberry	VS, VS				
Myrica pennsylvanica	Northern Bayberry				VT, T	Semi evergreen shrub. Full sun. Available
Nerium oleander	Oleander		T		VT, T	
Nyssa sylvatica	Black Tupelo				MT	
Olea europaea	Common Olive				T	
Ostrya japonica	Japanese Hophornbeam	MT				
Pachistina canbyi	Canby Pachistina				S	Suited only to protected, fertile sites. Acidic or neutral soils
Pachysandra terminalis	Japanese Spurge					
Parthenocissus gvinguefolia	Virginia Creeper	VS, MT	S		S	
Picea abies	Norway Spruce	VS	MT, MT		VS	
Picea glauca	White Spruce					

Table B-8 (Cont.).

LATIN NAME	COMMON NAME	Spray	Na Soil	Cl	General	COMMENTS
Picea pungens	Blue Spruce	VT	VS,MT		VS	
Pinus cembra	Swiss Stone Pine	VS				
Pinus divaricata	Jack Pine	T			T	
Pinus mugo mughus	Mugo Pine	T			T	
Pinus nigra	Austrian Pine	T				
Pinus ponderosa	Ponderosa Pine	VS	MT,MT		MT,T,T	
Pinus resinosa	Red Pine	VS			MT,VS	
Pinus strobus	White Pine	VS	S		VS	
Pinus sylvestris	Scotch Pine	VS,VS			MT	
Pinus thumbergi	Japanese Black Pine				VT	
Pittosporum spp	Pittosporum		MT		VS,MT	
Platanus acerifolia	London Planetree	S			S	
Populus species	Poplar	T	MT		T	
Populus acuminata	Lance Leaf Poplar				VT,T	
Populus alba	White Poplar	T	T			
Populus alba nirea	Silver Poplar	T	T		VT,VT	
(alba acerifolia)	Silver Poplar					
(alba arembergica)	Silver Poplar					
(alba argentea)	Silver Poplar					
Populus angustifolia	Narrow leaf Poplar				T	
Populus balsamifera	Southern Poplar				T	
(deltoides missoriensis)						
Populus canadensis	Carolina Poplar	VT				
Populus canescens	Gray Poplar	VT			VT,VT	
Populus deltoides	Eastern Poplar(Cottonwood)	VT	T		T	
Populus fremontii	Fremont Cottonwood				T	
Populus grandidentata	Largetooth Aspen	MT	MT		T	
Populus laurifolia	Laurel Poplar				VS,VS	
Populus nigra	Black Poplar	T			MT	

Table B-8 (*Cont*).

LATIN NAME	COMMON NAME	Spray	Soil Na	Soil Cl	General	COMMENTS
Populus nigra "Italica"	Lombardy Poplar	MT	VS		S,VS	
Populus sargestí	Plains Poplar				T	
Populus tremula	European Aspen				VT,S	
Populus tremuloider	Quaking Aspen	MT,MT			VT,ST	
Populus trichocarpa	California Poplar				T	
Potentila tridentata	Nineleaf Cinquefoil				S	Adapted to poor rocky and sandy soils. Good evergreen ground cover. Available
Prunus species	Prune	S			S	
Prunus armeniaca	Apricot	MT	T		VT,VT	
Prunus arium	Mazzard Cherry	VS			T	
Prunus mahaleb	Mahaleb Cherry					
Prunus maritima	Beach Plum	T			T	
Prunus padus	European Bird Cherry				MT,T	
Prunus persica	Peach	VS,S,VS			S	
Prunus serotina	Black Cherry	S			T	
Prunus spinosa	Blackthorn				T	
Prunus virginiana	Chokecherry	T				
Pseudotsuga menzesii	Douglas Fir	MT	VS,MT		VS	
Pseudotsuga taxifolia	Common Douglas Fir				VS,MT	
Puccinellia distans (L.)			MT		VT	
Pyracantha species	Pyracantha				S	
Pyracantha atalantoides	Gibbs Firethorn	VS				
Pyracantha coccinea	Scarlet Firethorn	VS				
Pyracantha crenatoserrata	Chinese Firethorn	VS				
Pyrus species	Pear	MT			S	
Pyrus baccata	Siberian Crab		MT		MT,MT	
Pyrus padus	European Bird Pear				MT	
Quercus species	Oak		VT,T		T,T	
Quercus alba	White Oak	VS	VT		VT,T,	
					VT	
Quercus bicolor	Swamp White Oak		VT		MT,T	
Quercus borealis	Northern Red Oak		VT		MT,T	

Table B-8 (Cont.).

LATIN NAME	COMMON NAME	Spray	Soil Na Cl	General	COMMENTS
Quercus cerris	Turkey Oak	S		MT,T	
Quercus coccinea	Scarlet Oak	S			
Quercus heterophylla	Bartram Oak	VS			
Quercus libani	Lebanon Oak	MT			
Quercus macranthera		MT,S		T	
Quercus macrocarpa	Bur Oak	VS			
Quercus muehlenbergi	Yellow Chestnut Oak	S		MT	
Quercus palustris	Pin Oak	VS			
Quercus petraea	Durmast Oak	VS			
Quercus pyrenaica	Pyrenees Oak				
Quercus robur	English Oak	S	T,VT	VT,VT	
Quercus rubra	Southern Red (Falcata)	S,T	T	VT,T	
	Eastern Red (Borealis maxima)	S,T	T	VT,T	
	Swamp Oak	S,T	T	T	
Quercus sessilifora	Durmast Oak)Petraca)			MT	
Rhamnus species	Buckhorn	MT,VT			
Rhamnus catharticus	European Buck				
Rhamnus crenatus	Oriental Buck	MT			
Rhamnus davurica	Dahurian Buckthorn	T			
Rhamnus frangula	Alder Buckthorn	MT,MT			
Rhamnus infectorious	Persianberry Buckthorn	MT,MT			
Rhamnus utilis	Chinese Buckthorn	VS			
Rhus aromatica	Fragrant Sumac			T	Dense growth. Accustomed to dry soils. Not commonly available
Rhus copallina	Flameleaf Sumac			VT	Woody, spreading shrub. Wide variety of soils. Has ability to occupy poor soil
Rhus glabra	Smooth Sumac			T	
Rhus glabra cismontana	Rocky Mountain Sumac		T,T	VT,VT, VT	Similar to but less hardy than Smooth Sumac
Rhus trilobata	Squawbush				

Table B-8 (Cont.).

LATIN NAME	COMMON NAME	Spray	Soil Na	Soil Cl	General	COMMENTS
Rhus typhina	Staghorn Sumac	T			T	
Ribes alpinum	Alpine currant	T, MT, T			T	
Ribes americanum	American Black Currant	MT				
Ribes aureum	Golden Currant	T				
Ribes diacanthum	Siberian Currant	T				
Ribes divaricatum	Straggly Gooseberry	MT				
Ribes giraldii		VS				
Ribes gardonianum	Gordon Currant	T				
Ribes Magdalenae		T				
Ribes migrum	European Black Currant	VS			MT, MT	
Ribes niveum	Snow Gooseberry	T				
Ribes sanguineum	Winter Currant	S				
Robina species	Locust	Locust			VT, VT, T	
Robinia lispida	Roseacacia	VT, VT			MT	
Robinia pseudoacacia	Black Locust		T		VT, VT, T	
Robinia pseudoacacia "Pyramidalis"	Pyramid Locust					
Rosa amblyotis	Kamchatka Rose	VT				Viborous suckering shrub rose. Not commercially available
Rosa canina		S, VS, S				
Rosa multiflora	Japanese Rose or Multiflora Rose		S, S, VS		VS, VS, MT	
Rosa nitida	Shining Rose				MT	
Rosa rubiginosa	Sweetbriar	VS			MT, T	
Rosa rugosa	Rugosa Rose	T	MT		MT, T	A scrambling, thorny plant. Adapted to many soils. Available from wild stock
Rosa wichuraiana	Wichura or Memorial Rose				T	
Rubus alleghenienisis	Allegheny Blackberry					Hardy native
Rubus flagellaris	Northern Dewberry				S	
Rubus fructicasus	European Blackberry	S, VS				
Rubus laciniatus	Cutleaf Blackberry				S	Inadequately hardy

Table B-8 (Cont.).

LATIN NAME	COMMON NAME	Spray	Soil Na	Cl	General	COMMENTS
Rubus parvifolius	Japanese Trailing Raspberry				T	Hardy ground cover. Soil must be fertile
Rubus rosendahli	Rosendahl Dewberry				S	Rare native; sandy, acidic soils, full sun
Salix alba	White Willow	T			T	
Salix alba calva	Pyramidal Weeping Willow				T	
Salix alba tristis	Golden Weeping Willow	MT			T	
Salix alba ritellina	Yellowstem Willow	VS	T,MT		T,T,VS	
Salix amygdalina	Peach-leaf Willow	T				
Salix aurita	Roundear Willow	MT				
Salix babylonica	Babylon Weeping Willow	MT,MT			VT,T	
Salix caprea	Great Willow	T			MT	
Salix daphnoides	Daphne Willow	T				
Salix fragilis	Brittle Willow	T				
Salix nigra	Black Willow					
Salix pentandra	Laurel Willow	MT				
Salix purpurea	Purple Osier Willow		T,VT		VS,VS,T	
Salix purpurea nana	Artic Blue Willow Blue Willow				VS,VS	
Salix purpurea lambertiana	Lambert Purpleasier Willow	T				
Salix tristis	Dwarf Gray or Pussy Willow				VS	Low shrubby native occurring on sandy acidic soils. Full sun.
Salix vitellina	Golden Willow	S			MT,T	
Sambucus canadensis	American Elder	S			MT	
Sambucus nigra	European Elder	VS,S,S			MT	
Sapirdus species	Soapberry				VT	
Sheperdia argentea	Silver Buffalo Berry	T,MT,T				
Sheperdia canadensis	Russet Buffalo Berry				T	
Sophora japenica	Japanese panoda Tree	VS				
Sorbus species	Mountain Ash	T				
Sorbus aria	White Beam Mountain Ash	T				

Table B-8 (Cont.).

LATIN NAME	COMMON NAME	Spray	Soil Na	Cl	General	COMMENTS
Rhus typhina	Staghorn Sumac	T				
Ribes alpinum	Alpine currant	T, MT, T			T	
Ribes americanum	American Black Currant	MT			T	
Ribes aureum	Golden Currant	T				
Ribes diacanthum	Siberian Currant	T				
Ribes divaricatum	Straggly Gooseberry	MT				
Ribes giraldii		VS				
Ribes gardonianum	Gordon Currant	T				
Ribes Magdalenae		T				
Ribes migrum	European Black Currant	VS			MT, MT	
Ribes niveum	Snow Gooseberry	T				
Ribes sanquineum	Winter Currant	S				
Robina species	Locust	Locust			VT, VT, T	
Robinia lispida	Roseacacia	VT, VT			MT	
Robinia pseudoacacia	Black Locust		T		VT, VT, T	
Robinia pseudoacacia "Pyramidalis"	Pyramid Locust	VT				
Rosa amblyotis	Kamchatka Rose					Viborous suckering shrub rose. Not commercially available
Rosa canina		S, VS, S			VS, VS,	
Rosa multiflora	Japanese Rose or Multiflora Rose		S, S, VS		MT	
Rosa nitida	Shining Rose				MT	
Rosa rubiginosa	Sweetbriar	VS			MT, T	
Rosa rugosa	Rugosa Rose	T	MT		MT, T	
Rosa wichuraiana	Wichura or Memorial Rose				T	
Rubus alleghenlensis	Allegheny Blackberry					A scrambling, thorny plant. Adapted to many soils. Available from wild stock
Rubus flagellaris	Northern Dewberry				S	Hardy native
Rubus fructicasus	European Blackberry	S, VS				
Rubus laciniatus	Cutleaf Blackberry				S	Inadequately hardy

Table B-8 (Cont.).

LATIN NAME	COMMON NAME	Spray	Na Soil	Cl	General	COMMENTS
Rubus parvifolius	Japanese Trailing Raspberry				T	Hardy ground cover. Soil must be fertile
Rubus rosendahli	Rosendahl Dewberry				S	Rare native; sandy, acidic soils, full sun
Salix alba	White Willow	T			T	
Salix alba calva	Pyramidal Weeping Willow				T	
Salix alba tristis	Golden Weeping Willow	MT			T	
Salix alba ritellina	Yellowstem Willow	VS			T, T, VS	
Salix amygdalina	Peach-leaf Willow	T	T, MT			
Salix aurita	Roundear Willow	MT				
Salix babylonica	Babylon Weeping Willow	MT, MT			VT, T	
Salix caprea	Great Willow	T			MT	
Salix daphnoides	Daphne Willow	T				
Salix fragilis	Brittle Willow	T				
Salix nigra	Black Willow					
Salix pentandra	Laurel Willow	MT				
Salix purpurea	Purple Osier Willow				VS, VS, T	
Salix purpurea nana	Artic Blue Willow Blue Willow				VS, VS	
Salix purpurea lambertiana	Lambert Purpleasier Willow	T	T, VT			
Salix tristis	Dwarf Gray or Pussy Willow				VS	Low shrubby native occurring on sandy acidic soils. Full sun.
Salix vitellina	Golden Willow	S			MT, T	
Sambucus canadensis	American Elder	S			MT	
Sambucus nigra	European Elder	VS, S, S			MT	
Sambucus racemosa						
Sapirdus species	Soapberry				VT	
Sheperdia argentea	Silver Buffalo Berry	T, MT, T				
Sheperdia canadensis	Russet Buffalo Berry				T	
Sophora japenica	Japanese panoda Tree	VS				
Sorbus species	Mountain Ash	T				
Sorbus aria	White Beam Mountain Ash	T				

Table B-8 (Cont.).

LATIN NAME	COMMON NAME	Spray	Na Soil	Cl	General	COMMENTS
Sorbus aucuparia	European Mountain Ash	VS				
Sorbus commixta	Korean Mountain Ash	MT			S	
Sorbus decora	Snowy Mountain Ash	T				
Sorbus hybrida	Oakleaf Mountain Ash	S				
Sorbus intermedia	Swedish Mountain Ash	VS				
Sorbus japonica	Japanese Mountain Ash	T				
Sorbus koehneana	Koehnes Mountain Ash	VS				
Sorbus latifolia		S				
Sorbus pohuashanensis		VS				
Sorbus rufo-ferruginea	Flameberry Mountain Ash	VS				
Sorbus sambucifolia	Siberian Mountain Ash	VS				
Sorbus serotina		T				
Sorbus torminalis	Wild Service Tree	S				
Sorbus vilmorinii	Vilmorin Mountain Ash	VS				
Spiraea species	Spirea Vanhouttii				VS	
Spiraea arguta compacta	Dwarf Garland Spirea					Needs fertile, cultivated site. Available
Spiraea billiardi	Billiard Spirea					Spreading shrub. Available
Spiraea x bumalda	Bulmalda Spirea	S				
Stephanandra incisa	Dwarf Cutleaf Stephanandra					Low, arching plant. Limited availability
Stewartia serrata		S				T	
Symphoricarpus albus	Snowberry	MT,T			T	Available. Not adequately tested
Symphoricarpus albus racemosus	Common Symphoricarpes				T,T	
Symphoricarpus x chenaulti	Chenault Coralberry	VS			S	Not hardy
Symphoricarpus x chenaulti, Hancock	Hancock Coralberry	VS			S	Not fully hardy
Symphoricarpus occidentalis	Wolfberry				T	Hardy. Easily propagated. Adapted to all soils. Full sun
Symphoricarpus orbiculatus	Coralberry	VS				
Syringa amurensis japonica	Japanese Tree Lilac	T				
Syringa vulgaris	Common Lilac	VS,T			VT,VT	
Tamarix species (pentandra)	Tamarix (5 Stamen Tamarix)	T		T,T	VT,VT	
Tamarix gallica	French Tamarix					

Table B-8 (Cont.).

LATIN NAME	COMMON NAME	Spray	Soil Na Cl	General	COMMENTS
Tamarix hispida	Kashgar Tamarix			VT,VT	
Tamarix pallasii				VT	
Taxus species	Yew	S			
Taxus baccata	English Yew	VS			
Thuja species	Arborvitae			T	
Thuja occidentalis	Eastern or American Arborvitae	S		T	
Thuja orientalis	Oriental Arborvitae			T	
Tilia species	Linden		MT	VS	
Tilia americana	Basswood	MT		VS	
Tilia cordata	Little Leaf Linden	T	VS,VS	MS,S,VS	
Tilia euchlora	Crimean Linden	S			
Tilia platyphylla	Bigleaf Linden	T			
Tsuga canadensis	Canada Hemlock	VS	VS	VS	
Tsuga heterophylla	Western Hemlock	MT			
Ulmus species	Elm			MT,S	
Ulmus americana	American Elm	T		T,VS	
Ulmus campestris	European Elm			VT,VT	
Ulmus (carpinifolia)	Smooth-leafed Elm	S,MT,Ml		VT,MT	
Ulmus glabra	Scotch Elm	T,T		T	
Ulmus hollandica major	Dutch Elm			MT	
Ulmus laevis	Russian Elm	MT			
Ulmus parrifolia	Chinese Elm			T	
Ulmus procera	English Elm			T	
Ulmus pumila	Siberian Elm	T		VT,T	
Ulmus pumila aborea	Narrow Siberian Elm	VS			
Vaccinium angustifolium	Lowbush Blueberry				Native low shrub. Hardy once established. Requires moisture and acidic soils. Available
Vaccinium corymbosum	Highbush Blueberry			VS	Not hardy. Acidic soils, full sun.
Vaccinium corrymbosum x angustifolium	Hybrid Blueberry			VS	Acidic soils, full sun. Limited value.

Table B-8 (Cont.).

LATIN NAME	COMMON NAME	Spray	Soil Na	Cl	General	COMMENTS
Vaccinium nurtilloides	Velvet Leaf Blueberry				MS	A native species on acidic soils. Too difficult to grow
Vaccinium vitis-idaea	Cowberry			VS,VS	T	
Viburnum species	Viburnums				MT	
Viburnum burkwosdi	Burkwood Viburnum				MT	
Viburnum carlesi	Fragrant Viburnum				MT	
Viburnum cassinoides	Witherod Viburnum				MT	
Viburnum dentatum	Arrowwood Viburnum				MT	
Viburnum lantana	Wayfaringtree Viburnum	S				
Viburnum lentago	Nannyberry Viburnum				MT	
Viburnum molle	Kentucky Viburnum	S				
Viburnum opulus	European Cranberry Bush	S,T				
Viburnum prunifolium	Blackhaw Viburnum				S	
Viburnum sieboldi	Siebold Viburnum				MT	
Viburnum tomentosum	Doublefile Viburnum				MT	
Vinca major	Greater Periwinkle					
Vinca minor	Common or Lesser Periwinkle				MT	
Weigela 'Eva Rathke'	'Eva Rathke' Weigela		MT			
Xanthorhiza simplicissima	Yellowroot				VS	Needs moist soil. Acidic or neutral soils
Xylosma congestum	Xylosma		MT			
Yucca filamentosa	Adams Needle		MT			
Zelkova serrata	Japanese Zelkova				MT	

Table B-9. Freezing Point of Magnesium Chloride Solutions

MgCl$_2$ % by weight	Specific gravity @ 60°F (15.5°C)	Freezing point		Weights of solution and solute			
		°C	°F	MgCl$_2$ (lb/gal)	Brine (lb/gal)	MgCl$_2$ (kg/L)	Brine (kg/L)
10	1.086	−7.8	17.9	0.906	9.06	0.11	1.09
20	1.180	−27.4	−17.3	1.97	9.84	0.24	1.18
21	1.190	−30.6	−23.0	2.08	9.93	0.25	1.19
22	1.200	−32.8	−27.0	2.20	10.0	0.26	1.20
23	1.210	−28.9	−20.0	2.32	10.1	0.28	1.21
24	1.220	−25.6	−14.0	2.44	10.2	0.29	1.22
25	1.230	−23.3	−10.0	2.57	10.3	0.31	1.23
26	1.241	−21.1	−6.0	2.69	10.4	0.32	1.25
27	1.251	−19.4	−3.0	2.92	10.4	0.35	1.25
28	1.262	−18.3	−1.0	2.95	10.5	0.35	1.26
29	1.273	−17.2	+1.0	3.08	10.6	0.37	1.27
30	1.283	−16.7	+2.0	3.21	10.7	0.38	1.28

cial uses are for dust control, for water treatment, as refrigerator brines, as fire-extinguishing agents, and for fireproofing wood. The use of magnesium chloride in liquid form for ice control is increasing, and in that form the chemical is completely dissociated, so the only relevant value is the concentration of MgCl$_2$. Eutectic temperature is −33°C (−27°F) at a concentration of 21.6 percent. Proprietary mixtures containing 20 to 25 percent MgCl$_2$ along with various corrosion-inhibiting additives are available. One proprietary compound (Freezgard® + PCI®) reportedly has a eutectic temperature of −20°C (−4°F). These solutions are effective ice-melting agents at temperatures above −7°C (19°F).

Source

Though some of this ice control chemical is synthesized, the principal sources are brines from the Great Salt Lake in the United States and the Dead Sea in Israel. It is available in solid (flake) form as the hexahydrate (MgCl$_2$•6H$_2$O). Some vendors appear to state that their product contains the anhydrous form, MgCl$_2$ (that is, with no water of crystallization). The anhydrous product cannot be made simply by heating the hexahydrate. Doing so will produce hydrogen chloride (HCl) and magnesium oxide. The important information to look for is the percentage of magnesium chloride.

Calcium Magnesium Acetate $(CaMg_2(CH_3COO)_2)_6$

Commonly called CMA, calcium magnesium acetate is not a naturally occurring chemical but was first synthesized in 1980 during research funded by the Federal Highway Administration that was aimed at development of a noncorrosive ice control chemical. CMA is primarily a mixture of calcium and magnesium acetates; a 3/7 Ca/Mg molar ratio was found to be optimum in the FHWA studies. The eutectic temperature for this composition is about $-18°F$ ($-28°C$) at a concentration of 32.6 wt%.

Source

Currently there is only one commercial method for manufacturing CMA: reacting acetic acid with dolomitic limestone. The acetic acid, currently manufactured from natural gas or petroleum, is the costly component of the compound. Experimental quantities have been produced by biodegradation of agricultural wastes, and research is under way to develop other low-cost methods of producing acetic acid. The compound is available as pellets and as a liquid.

Effectiveness

CMA is not highly effective as a deicing chemical in solid form because of its affinity for water and its low particle density, which reduces its ability to penetrate a snowpack to reach the pavement. One of its benefits, however, is its capacity for keeping snow "mealy," thereby reducing the likelihood it will compact.

Though CMA is not as soluble in water as NaCl and $CaCl_2$, a solution can be made at point of use to serve as a prewetting agent or as a straight chemical application. A solution of nearly 25 percent is made by mixing 2000 lb of calcium magnesium acetate in 720 gal of water. Little information concerning its effectiveness and economics has been gained as yet.

Urea, NH_2CONH_2

Source

Urea is made by reacting anhydrous ammonia and carbon dioxide at high pressure and at temperatures of 180 to 200°C (356 to 392°F). It is produced in large quantities principally for the fertilizer market.

Though it is available commercially in liquid form in 50 to 80 percent concentrations, the solid form in fertilizer grade has been used as an ice control chemical. It first received widespread use in that role on airfields because of its low corrosion potential and the prohibition of chloride chemicals in those areas.

Environmental Impact

Ammonia, which is highly toxic to aquatic organisms, is released when urea breaks down. Its high nitrogen content can accelerate the formation of algae blooms, and its high biological oxygen demand (BOD) can deplete the oxygen essential for survival of fish and other aquatic animals. A study by Transport Canada concluded that it is impractical to mitigate the impacts of urea use, and so alternative ice control chemicals should be used.

Material Standards

Urea for use on airfields must meet the SAE specification AMS 1431A, *Compound, Solid Runway and Taxiway Deicing/Anti-icing,* and the military specification MIL SPEC DOD-U-10866D, *Urea–Technical.* Either powder or prilled forms are used. [Prilled, or "shotted," form refers to the spherical shape of the particles, with diameter of about $\frac{1}{16}$ in. (1.5 mm)].

Effectiveness

Urea is typically applied at the same rate as sodium chloride over the same temperature range, although its melting action is slightly slower than that of sodium chloride.

Ethylene Glycol, $C_2H_6O_2$

This chemical is familiar to motorists, as it is commonly used as an automobile radiator antifreeze. This is the result of its mixing completely with water, low freezing point, stability, and low rate of corrosion. These properties have also led to its use in large quantities on airfields, but its toxicity and high biological oxygen demand have resulted in decreased use on runways, although it remains the principal component of the Type I and II aircraft deicing and anti-icing fluids (see Sec. 11.3).

Appendix B

Propylene Glycol, $C_3H_8O_2$

Ethylene glycol is highly toxic to aquatic and mammalian organisms. For this reason, it is being replaced by the less toxic propylene glycol for aircraft deicing. However, while ethylene glycol and propylene glycol are both biodegradable, propylene glycol degrades at a slower rate and has a greater BOD. As a consequence, it will remain in the environment longer than ethylene glycol and will consume more oxygen while breaking down. It can still be harmful to the environment.

Potassium Acetate, $KC_2H_3O_2$

This chemical, commonly referred to as KAc, is manufactured by the reaction of acetic acid with potassium carbonate (potash). KAc is a white, crystalline, deliquescent powder with a salty taste. It is gaining acceptance for application on airport runways as a 50 percent concentration liquid as a consequence of its low corrosion potential and melting effectiveness.

Sodium Acetate, $NaC_2H_3O_2$

The search for ice control chemicals that are less corrosive than chlorides has been driven by the requirements of the aviation industry. Use of corrosive chemicals is prohibited on all aircraft operational areas. Sodium acetate is another in a series of chemicals that began with calcium magnesium acetate whose effectiveness and low corrosion meet airfield requirements.

Sodium Formate, HCOONa

Because of its low corrosion rate, sodium formate has been investigated for use on airfields since the mid-1980s. Its use in that application is increasing, although cost has limited its use on highways. It is nearly as effective as sodium chloride in its melting effectiveness, and it retains effectiveness to a lower temperature than NaCl.

Source

Sodium formate is a white crystalline salt prepared by passing carbon monoxide through heated sodium hydroxide (caustic soda). It is also produced as a coproduct of the manufacture of polyol chemicals such as pentaerythritol.

Ice Control Chemical Properties

Many chemicals have been investigated for use as ice control agents. Factors that need to be considered for a material suggested for use in this application include

- Extent to which it lowers the freezing point of water
- Rate at which it dissolves (solubility)
- Effects on materials of construction (corrosion, spalling)
- Effects on the environment (killing of plants and wildlife)
- Toxicity (how much is required to cause death)
- Cost and availability

The performance required is so stringent that few of the many thousands of chemical compounds meet the test. The following tables present the properties of ice control chemicals currently in use, and also of some that have been used in the past but have now fallen into disfavor and some that are still undergoing evaluation. The latter two classes are included for comparison purposes only. The values for the pure compounds are taken from the literature; commercial products that are predominantly the listed chemical may contain some additives, but the values given will not vary to an appreciable extent and may be used for any calculations. The many combinations of the listed chemicals sold as commercial products are not included. The properties of combinations cannot in most cases be predicted from knowledge of the composition and proportions. Suppliers of the materials should be contacted for this information.

The chemicals included in these tables are

Calcium chloride

Calcium magnesium acetate

Ethylene glycol

Isopropyl alcohol

Magnesium chloride

Potassium acetate

Appendix B

Potassium chloride
Propylene glycol
Sodium acetate
Sodium chloride
Sodium formate
Urea

Explanation of Terms in the Tables

Heading	Chemical name of compound.
Composition	Chemical formula.
Alternative names	Common, chemical, or commercial names.
CAS	The Chemical Abstracts Service Registry Number, a unique number assigned to substances recorded in the Chemical Abstracts Service Registry System of the American Chemical Society.
RTECS	Registry of Toxic Effects of Chemical Substances, a comprehensive database of basic toxicity information for over 100,000 chemical substances prepared by the U.S. National Institute for Occupational Safety and Health (NIOSH).
Form	Liquid or solid, appearance (most common form).
Eutectic temperature/ composition	The temperature at which a chemical solution freezes completely without change in composition.
Lowest effective temperature	The lowest temperature at which a chemical should be applied for effective, timely melting action.
Solubility	The amount of the chemical that will dissolve in a specified amount of water at a specific temperature, usually stated as number of grams of the chemical per 100 g water.
Melting effectiveness	Qualitative expression of the extent of melting produced by the chemical.

Corrosion	Brief comment on the extent of corrosion produced by the chemical, or other effects the chemical may have on materials such as portland cement concrete.
Surface effects	The condition of the pavement surface produced by the chemical, e.g., slippery.
Moisture absorption	Whether the chemical absorbs moisture (hygroscopic) or dissolves (deliquescent).
Toxicity	The harmful level in terms of LD_{50}, the dose that kills 50 percent of test animals, usually rats fed the chemical by mouth. The dose is expressed as the amount (in milligrams) for each kilogram of body weight. For example, in human terms, 50 percent of people weighing 70 kg (154 lb) who swallowed 210 g (about 1/2 lb) of salt would be killed (based on an LD_{50} of 3000 mg/kg). *Note:* Only LD_{50} values are listed; many other toxicological properties, such as carcinogenicity, teratogenicity, neurotoxicity, oncogenicity, and toxic levels for aquatic biota, can be found in the Material Safety Data Sheet (MSDS) from vendors. LDLo is Lethal Dose low, the lowest dose which causes death in test animals.
Effects on humans	Other, nonlethal, effects that the chemical may produce on humans.
Storage and handling	Things to consider in storing and handling the chemical.
Environmental effects	What the chemical may do to the environment.
Cost	Price per unit weight in U.S. dollars (from *Chemical Marketing Reporter*, when known).

Sources of Information in the Table are Identified by a Letter:

C	Cryotech Deicing Technology, Fort Madison, Iowa 52627
E	Eastman Chemical Products, Inc., Kingsport, Tenn. 37662
F	Fisher Scientific, 1 Reagent Lane, Fair Lawn, NJ 07410

Appendix B

H	Hoechst Canada, 800 René-Lévesque Blvd. West, Montreal, QC Canada H3B 1Z1
K	Kirk-Othmer (Howe-Grant 1992)
L	*Lange's Handbook of Chemistry* (Dean 1992)
M	*The Merck Index* (Merck 1989)
S	Sigma-Aldrich (Lenga 1988)
V	Vershueren (1996)

Calcium Chloride

Composition	$CaCl_2$
Alternative names	Calcium chloride
	CAS: 10043-52-4 RTECS: EV9800000
Form	Solid; available as flakes (73 to 80 percent $CaCl_2$) or pellets (95 to 98 percent $CaCl_2$). Liquid $CaCl_2$ is also available in concentrations of 32 percent $CaCl_2$ and higher.
Eutectic temperature/composition	−51°C (−60°F) @ 29.8 wt%
Lowest effective temperature	−29°C (−20°F)
Solubility	977 g/L (0°C) (dihydrate) 3.3 kg/L (60°C [140°F])
Melting effectiveness	Liberates heat when it goes into solution, aiding in its high melting effectiveness.
Corrosion	Aggressively promotes corrosion of ferrous metals.
Surface effects	Leaves moist film on road which reduces friction slightly.
Moisture absorption	Very hygroscopic; readily absorbs moisture from the air and liquefies completely.
Toxicity	Oral rat LD_{50}: 1000 mg/kg (S).
Effects on humans	Not toxic in ordinary amounts workers are exposed to; it is used medically as an electrolyte replacement and as a diuretic.
Storage and handling	Solid form must be protected from the atmosphere to prevent absorption of water vapor.
Environmental effects	Though the calcium in this compound is more beneficial to soil than sodium from salt, it adds hardness to water. BOD: 0.
Cost	80 percent flake: ~$250/ton
	94–97 percent flake (bulk): $275/ton

Calcium Magnesium Acetate (CMA)

Composition	$CaMg_2(CH_3COO_2)_6$
Alternative names	CAS: 76123-46-1
Form	Solid; white pellets or granules
Eutectic temperature/composition	−27.5°C (−17.5°F) at 3:7 Ca:Mg ratio @ 32.5 wt% (K)
Lowest effective temperature	−7°C (20°F)
Solubility	
Melting effectiveness	Does not perform well as a deicing agent—particles float on the surface of snow or ice—but acts most effectively to keep snow mealy and easy to plow. Requires about 40 percent more material to equal salt's effectiveness.
Corrosion	Does not promote corrosion, but rather acts as a corrosion inhibitor.
Surface effects	Not aggressive to concrete; has a residual effect.
Moisture absorption	No significant moisture absorption.
Toxicity	Oral rat LD_{50}: >5000 mg/kg (McFarland et al. 1992) (C).
Effects on humans	Not hazardous, though product may have the odor of vinegar.
Storage and handling	No personnel problems in handling; solid material should be protected from the weather.
Environmental effects	Acetate ion may contribute to loss of oxygen in water bodies.
Cost	$650/ton

Ethylene Glycol

Composition	$C_2H_6O_2$
Alternative names	1,2-ethanediol; 1,2-dihydroxyethane; 2-hydroxyethanol CAS: 107-21-1 RTECS: KW2975000
Form	Colorless, hygroscopic liquid; mild odor.
Eutectic temperature/composition	−50°C (−58°F) @ abt wt %
Lowest effective temperature	−50°C (−58°F)

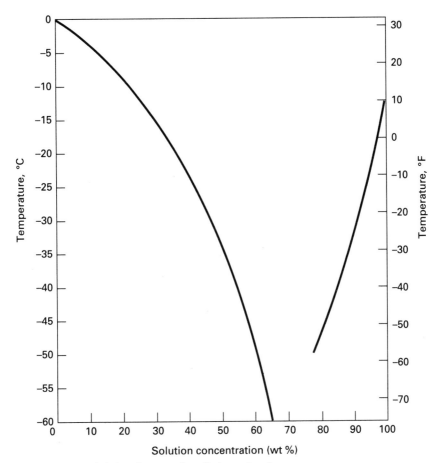

Figure B-3. Solubility diagram for ethylene glycol.

Solubility	≥100 mg/mL @ 17°C (63°F).
Melting effectiveness	Excellent.
Corrosion	Not corrosive.
Surface effects	Leaves slightly slippery surface.
Moisture absorption	Hygroscopic—absorbs twice its weight of water at 100 percent relative humidity.
Toxicity	Oral rat LD_{50}; 4700 mg/kg. Oral human LDLo: 786 mg/kg. Lethal human dose: 1.4 mL/kg (K).
Effects on humans	Harmful or fatal if swallowed; a teratogen.

Storage and handling	Requires protective clothing and breathing equipment; TLV 50 ppm vapor (K).
Environmental effects	BOD_5: 13 to 54 (V); BOD_5: 0.16 to 0.81 g O_2/g (E); BOD: 1500 mg O_2/g (H); COD: 95 to 98 (V). Daphnia magna 24-h LC_{50}: 74 g/L; 48-h LC_{50}: >41 g/L (E).
Cost	N.A. [Not Available]
Comments	Ethylene glycol has been used as spot treatment on bridge decks and on airport runways and ramps. Cost, safety, and environmental concerns have resulted in reduced use. The chemical is the principal component of aircraft deicing fluids.

Isopropyl Alcohol

Composition	C_3H_8O
Alternative names	2-propanol, isopropanol, dihydrocarbinol, dimethylcarbinol; sec-propyl alcohol
	CAS: 67-63-0 RTECS: NT8050000
Form	Flammable liquid.
Eutectic temperature/composition	Freezing point (100 percent) $-89.5°C$ ($-129°F$)
Lowest effective temperature	$-50°C$ ($-58°F$)
Solubility	Very soluble in water (>1000 mg/L); insoluble in salt solutions.
Melting effectiveness	Excellent melting action.
Corrosion	Noncorrosive.
Surface effects	No significant effect.
Moisture absorption	No data.
Toxicity	Oral rat LD_{50}: 5.8 g/kg (M). Lethal human dose: 100 mL (M).
Effects on humans	Ingestion or inhalation of large quantities of vapor may cause flushing, headache, dizziness, mental depression, nausea, vomiting, narcosis, anesthesia, coma (M).
Storage and handling	Avoid breathing large quantities of vapor.
Environmental effects	Daphnia magna 24-h LC_{50}: 9500 mg/L. BOD: 2.4 g $O_2/g(5)$.
Cost	N.A.

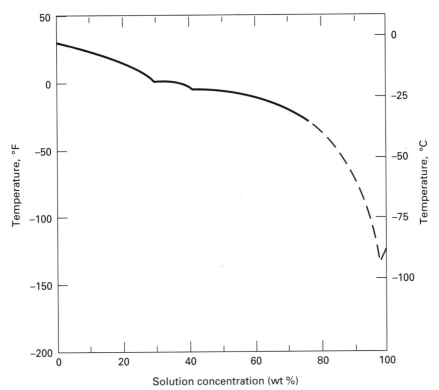

Figure B-4. Solubility diagram for isopropyl alcohol.

Magnesium Chloride

Composition/Alternative names	$MgCl_2 \cdot 6H_2O$ CAS: 7791-18-6 RTECS: OM2975000
	$MgCl_2$ CAS: 7786-30-3 RTECS: OM2800000
Form	Solid: white crystals, deliquescent, hexahydrate form (46 percent $MgCl_2$). Liquid: about 30 percent solution used as an ice control chemical.
Eutectic temperature/composition	−33°C (−27°F) at 21.6 wt%
Lowest effective temperature	−23°C (−10°F).
Solubility	1 g dissolves in 0.6 ml water (M). 35.3 g/100 g @ 20°C (68°F) (L).
Melting effectiveness	Very similar to $CaCl_2$ in its performance, though it has slightly less melting capability.

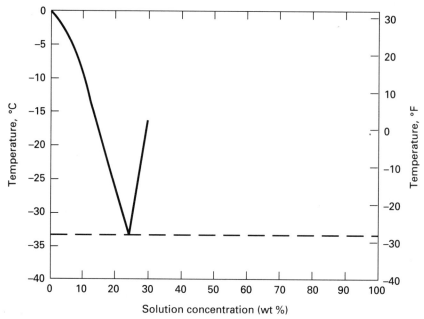

Figure B-5. Solubility diagram for magnesium chloride.

Corrosion	Slightly corrosive; low potential for spalling concrete.
Surface effects	Has demonstrated a significant residual effect on pavement without producing a slippery film.
Moisture absorption	Solid form readily absorbs moisture from the air and liquefies (it is deliquescent).
Toxicity	Oral rat LD_{50}: 2800 mg/kg. For hexahydrate, oral rat LD_{50}: 8100 mg/kg (M).
Effects on humans	Very low toxicity, though may cause slight eye, nose, and skin irritation.
Storage and handling	No special handling by personnel required. Dry chemical must be protected from moisture in air to prevent caking or liquefying.
Environmental effects	In large quantities can contribute to eutrophication of water bodies.
Cost	$50/ton

Appendix B

Potassium Acetate

Composition	CH_3COOK
Alternative names	Potassium ethanoate; Clearway 1; Cryotech E36; Safeway KA; Octagon RD 1435
	CAS: 127-08-2 RTECS: AJ3325000
Form	Colorless deliquescent crystals or white crystalline flakes. Available commercially as a liquid of 50 percent concentration.
Eutectic temperature/composition	−60°C (−76°F) @ 50 wt%
Lowest effective temperature	−25°C (−13°F)
Solubility	255 g/100 g H_2O (20°C) (L); 2 g/mL H_2O (M)
Melting effectiveness	Excellent.
Corrosion	Low; commercial product is inhibited.
Surface effects	Slightly slippery film.
Moisture absorption	Absorbs moisture from air and dissolves.

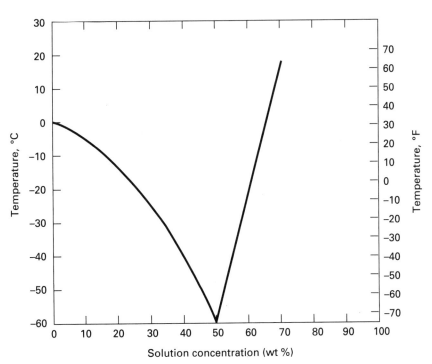

Figure B-6. Solubility diagram for potassium acetate.

Toxicity	Daphnia magna 48-h LC_{50}: >3000 mg/L (C). Fathead minnow 7-day LC_{50}: >1500 mg/L. Oral rat LD_{50}: 3250 mg/kg (F).
Effects on humans	Nonhazardous.
Storage and handling	Solid form must be protected from moisture.
Environmental effects	BOD: 300 mg O_2/g (H). BOD_5: 140 mg O_2/g @ 20°C (C)*Cost*

Potassium Chloride

Composition	KCl
Alternative names	Muriate of potash
	CAS: 7447-40-7 RTECS: TS8050000

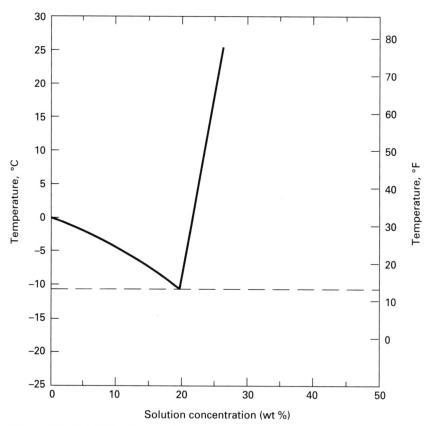

Figure B-7. Solubility diagram for potassium chloride.

Appendix B

Form	Solid; white crystals or crystalline powder.
Eutectic temperature/composition	−10.7°C (12.7°F) @ 19.9 wt% (SE)
Lowest effective temperature	−5°C (23°F)
Solubility	22 g/100 g H_2O (0°C) (32°F). 25.8 g/100 g H_2O (25°C) (77°F) (L).
Melting effectiveness	Similar to NaCl.
Corrosion	Corrodes ferrous metals.
Surface effects	Similar to NaCl.
Moisture absorption	Hygroscopic
Toxicity	Oral rat LD_{50}: 2600 mg/kg (S). Oral human LDlo: 20 mg/kg (S).
Effects on humans	Irritant; do not inhale.
Storage and handling	Keep dry
Environmental effects	Benign; used as a fertilizer. BOD:O
Cost	$125/ton

Propylene Glycol

Composition	$C_3H_8O_2$ $HOC_2H_3(CH_3)OH$
Alternative names	1,2-propanediol; 1,2-dihydroxypropane; methyl glycol
	CAS: 57-55-6 RTECS: TY2000000
Form	Hygroscopic, viscous colorless liquid.
Eutectic temperature/composition	−60°C (−76°F)
Lowest effective temperature	−25°C (−11°F)
Solubility	Completely miscible with water.
Melting effectiveness	Excellent.
Corrosion	Very low.
Surface effects	Slippery film.
Moisture absorption	Hygroscopic
Toxicity	Oral rat LD_{50}: 25 mL/kg (M).
Effects on humans	Relatively harmless; used in foods, drugs, cosmetics.
Storage and handling	Protect from air.
Environmental effects	At >10,000 mg/L, no effect on aquatic life (K). BOD_5: (1.08g O_2/g (E).

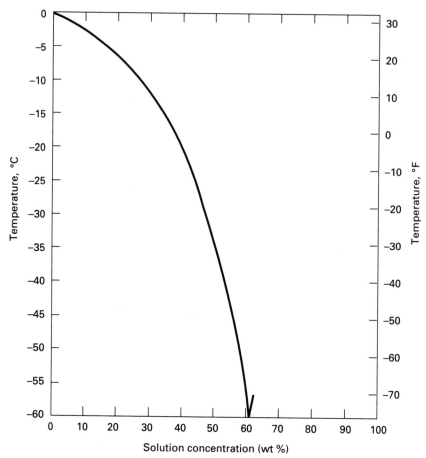

Figure B-8. Solubility diagram for propylene glycol.

Sodium Acetate

Composition	CH_3COONa
Alternative names	Safeway SD (Hoechst), NAAC (Cryotech)
	CAS: 127-09-3
Form	Hygroscopic powder; trihydrate transparent crystals.
Eutectic temperature/composition	−17°C (2°F) @ 23 wt% (H)
Lowest effective temperature	−10°C (14°F)
Solubility	365 g/L @ 20°C exothermic (H)

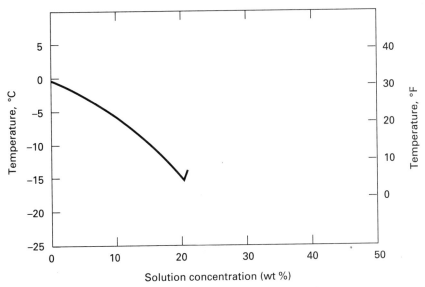

Figure B-9. Solubility diagram for sodium acetate.

Melting effectiveness	Good; heat produced in dissolving
Corrosion	Noncorrosive.
Surface effects	Does not produce slippery surface.
Moisture absorption	Hygroscopic.
Toxicity	Oral rat LD_{50}: 3530 mg/kg (S).
Effects on humans	Mildly toxic (ITI).
Storage and handling	Ordinary precautions, similar to salt.
Environmental effects	BOD: 780 mg O_2/g (H). Daphnia magna 48-h LC_{50}: >1000 mg/L (H).
Cost	N.A.

Sodium Chloride

Composition	NaCl
Alternative names	Salt
	CAS: 7647-14-5 RTECS: VZ4725000
Form	Solid colorless crystals or white granules; occurs in nature as the mineral halite and as a constituent of sea water.
Eutectic temperature/composition	−21°C (−5.8°F) at 23.3 wt%
Lowest effective temperature	−7°C (20°F)

Solubility	35.7 g/100 g @ 0°C; 97 g/100 g @ 20°C (see Fig. B-2).
Melting effectiveness	Very effective.
Corrosion	Promotes active corrosion of ferrous metals in the presence of air and moisture.
Surface effects	Dries completely, leaving a white residue. Causes spalling of portalnd cement concrete.
Moisture absorption	Pure salt does not absorb moisture; moisture will be absorbed if magnesium and calcium chlorides are present as impurities.
Toxicity	Oral rat LD_{50}: 3000 mg/kg; skin rabbit: 500 mg/24 h mild; eye rabbit: 10 mg/24 h moderate; indicates human toxic dose of 200 to 280 g (7 to 10 oz) for a 70-kg (155-lb) adult.
Effects on humans	Not considered toxic. However, salt is a source of sodium, which may contribute to high blood pressure in many people.
Storage and handling	Stockpiled material will form hard crust if exposed to rain or snow. Should be kept under cover on an impermeable base to prevent leaching into soil. Will absorb very little moisture from the air. No special precautions are required for personnel handling salt.
Environmental effects	High concentrations cause plant stresses and in the aquatic environment. BOD: 0.
Cost	$20 to 35/ton

Sodium Formate

Composition	HCOONa
Alternative names	Sodium methanoate; formic acid sodium salt; Salachlor; Octagon RD 1431SF; Safeway©SF
	CAS: 141-53-7 RTECS: LR0350000
Form	Solid: white, odorless, deliquescent crystalline granules.
Eutectic temperature/composition	−18°C (0°F) at 25 wt%
Lowest effective temperature	−11°C (12°F)

Appendix B

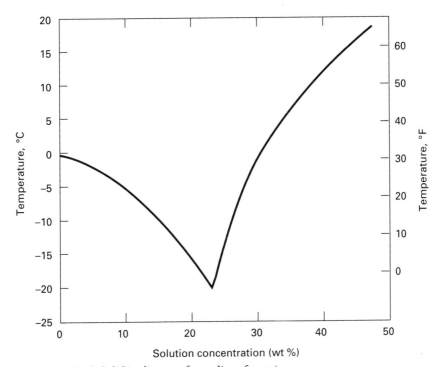

Figure B-10. Solubility diagram for sodium formate.

Solubility	44 lb/100 lb water @ 0°C; 97.2 lb/100 lb water @ 20°C (H); 365 g/L @ 20°C (H).
Melting effectiveness	More effective than urea and CMA and almost as effective as NaCl.
Corrosion	Noncorrosive to most metals; significantly less so than sodium chloride.
Surface effects	Similar to sodium chloride in surface degradation of portland cement concrete at same concentrations.
Moisture absorption	No data
Toxicity	Oral mouse LD_{50}: 11,200 mg/kg (F). Acute inhalation LC_{50}: >670 mg/m^3 (H).
Effects on humans	No serious effects
Storage and handling	Ordinary precautions; avoid breathing dust.
Environmental effects	Daphnia magna 48-h LC_{50}: >1000 mg/L. BOD: 230 mg O_2/g (H).
Cost	N.A.

Urea

Composition	$(NH_2)_2CO$
Alternative names	Carbamide; carbonyldiamide
	CAS: 57-13-6 RTECS: YR6250000
Form	Solid. White granular pellets which dissolve completely in water. Common use is as a fertilizer.
Eutectic temperature/composition	−11.7°C (11°F) at 32.6 wt%
Lowest effective temperature	−4°C (24°F)
Solubility	Dissolves readily in water, 1 g/mL H_2O (M).
Melting effectiveness	Not highly effective as an ice control chemical. It is three to four times slower than salt.

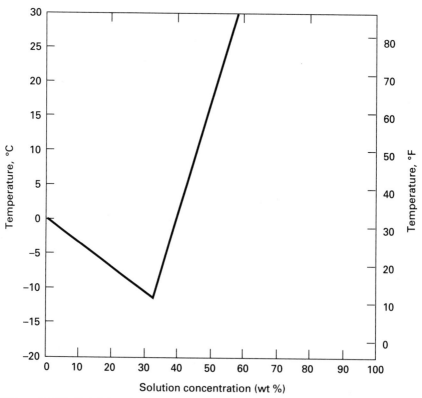

Figure B-11. Solubility diagram for urea.

Appendix B

Corrosion	Some evidence of corrosion of silver- and copper-containing metals, but no significant corrosion of ferrous metals.
Surface effects	Produces slippery film on pavement.
Moisture absorption	Very hygroscopic (absorbs moisture) at relative humidities above 80 percent at temperatures above 15°C (59°F).
Toxicity	Oral rat LD_{50}: 14,300 mg/kg (S).
Effects on humans	Nontoxic. It is an antiseptic which has been used to promote healing in infected wounds. However, it can irritate the skin, eyes, nose, and throat, so exposure to the dust should be avoided. Skin irritation: 22 mg/3 days was mild (S). Can develop the odor of NH_3 (ammonia).
Storage and handling	Breathing of the dust should be avoided.
Environmental effects	BOD: 2100 mg O_2/g (H). High nitrogen content has promoted eutrophication of lakes (algae blooms, loss of fish populations). Dissociates in water bodies into ammonia, which can kill fish.
Cost	$120/ton

References

Amundson, W. H. (1985). Presentation at University of Wisconsin (Madison) Snow and Ice Control Workshop, October 17–18.

Anon. (1971). *Environmental Impact of Highway Deicing*. Water Pollution Control Research Series 11040 GKK. Edison, N.J.: Edison Water Quality Laboratory.

Anon. (1971a). *Ideas & Actions: A History of the Highway Research Board*. Washington, D.C.: Highway Research Board.

Anon. (1995). *An Evaluation of Winter Runway Friction Measurement Equipment, Procedures, and Research*. Winter Runway Friction Measurement and Reporting Working Group (Federal Aviation Administration, Transport Canada, National Aeronautics and Space Administration, Airports Council International, American Association of Airport Executives, Regional Airline Association, Air Transport Association, and Air Line Pilots Association). Available online at http://www.bts.gov/ntl/data/wrfmrwg1.pdf

Bader, Henri (1962). *The Physics and Mechanics of Snow as a Material*. Cold Regions Science and Engineering Monograph II-B. Hanover, N.H.: U.S. Army Cold Regions Research and Engineering Laboratory.

Besselievre, William C. (1976). *Automatic Controllers for Hydraulically Powered Deicing-Chemical Spreaders*. Report No. FHWD-RD-76-505. Washington, D.C.: Federal Highway Administration.

Best, Gerald M. (1966). *Snowplow. Clearing Mountain Rails*. Berkeley, Calif.: Howell-North Books.

Browne, F. P., T. D. Larson, P. D. Cady, and N. B. Bolling (1970). *Deicer Scaling in Concrete*. University Park: Department of Civil Engineering, The Pennsylvania State University.

BWI (1993). *Aircraft Deicing Plan, Baltimore Washington International Airport*, p. 14 (November).

Chance, Robert L. (1974). Corrosion, deicing salts, and the environment. *Materials Performance*, 13(10), 16–22.

Chi, S. W. (1976). *Heat Pipe Theory and Practice: A Sourcebook*. Washington, D.C.: Hemisphere Publishing Corp.

Chisholm, D., and C. Eng (1971). *The Heat Pipe*. M&B Technical Library, TL/ME/2. London: Mills & Boon Limited.

Code of Federal Regulations. 40 CFR Sec. 50.6, National primary and secondary ambient air quality standards for particulate matter. Washington, D.C.: U.S. Government Printing Office.

Colbeck, Samuel C. (1978). *The Compression of Wet Snow*. Report 78-10. Hanover, N.H.: U.S. Army Cold Regions Research and Engineering Laboratory.

Conner, Billy, and Richard Gaffi (1982). *Optimum Sand Specifications for Roadway Ice Control*. Report No. FHWA-AK-RD-82-26. Fairbanks: Alaska Department of Transportation and Public Facilities.

Dean, John A. (1992). *Lange's Handbook of Chemistry*. 14th ed. New York: McGraw-Hill

Doesken, Nolan J., and Arthur Judson (1996). *The Snow Booklet: A Guide to the Science, Climatology, and Measurement of Snow in the United States*. Fort Collins: Colorado State University, pp. 9–10.

Dolce, John E. (1992). *Fleet Manager's Guide to Specification and Procurement*. Warrendale, Penn.: Society of Automotive Engineers, Inc.

Dunn, Peter D., and D. A. Reay (1976). *Heat Pipes*. New York: Pergamon Press.

FAA (1989). *Airport Design*. AC 150/5300-13. (9-29-89); and change 4, 11-10-94. Washington, D.C.: Federal Aviation Administration.

FAA (1991). *Airport Winter Safety and Operations*. AC 150/5200-30A (10-1-91). Change 1(11-22-91). Change 2(3-27-95). Washington, D.C.: Federal Aviation Administration.

FAA (1991a). *Runway Surface Condition Sensor Specification Guide*. AC 150/5220-13B (3-27-91). Washington, D. C.: Federal Aviation Administration.

FAA (1992). *Buildings for Storage and Maintenance of Airport Snow and Ice Control Equipment and Materials*. AC 150/5220-18 (10-15-92). Washington, D.C.: Federal Aviation Administration.

FAA (1992a). *Airport Snow and Ice Control Equipment*. AC 150/5220-20 (6-30-92). Washington, D.C.: Federal Aviation Administration.

FAA (1993). *Design of Aircraft Deicing Facilities*. AC 150/5300-14 (8-23-93). Washington, D.C.: Federal Aviation Administration.

FAA (1997). *Measurement, Construction, and Maintenance of Skid Resistant Airport Pavement Surfaces*. AC 150/5320-12C (3-18-97). Washington, D.C.: Federal Aviation Administration.

Finch, James W. (1993). *Motor Truck Engineering Handbook*. Warrendale, Penn.: Society of Automotive Engineers, Inc.

Fitzpatrick, Michael W., and David A. Law (1978). *Automatic Controls for Salt and Abrasive Spreaders*. Transportation Research Record 674, 64-7. Washington, D.C.: Transportation Research Board.

Franks, Felix, ed. (1972). *Water: A Comprehensive Treatise*. Vol. 1, *The Physics and Physical Chemistry of Water*. New York: Plenum Press.

Geer, Ira W., ed. (1996). *Glossary of Weather and Climate with Related Oceanic and Hydrologic Terms*. Boston, Mass.: American Meteorological Society.

Gerdel, R. W. (1954). The transmission of water through snow. *Trans. Am. Geophys. Union* 35(3), 475–485.

References

Glen, John W. (1974). *The Physics of Ice.* Cold Regions Science and Engineering Monograph II-C2a. Hanover, N.H.: U.S. Army Cold Regions Research and Engineering Laboratory.

Hanes, R. E., L. W. Zelazny, K. G. Verghese, R. P. Bosshart, E. W. Carson Jr., R. E. Blaser, and D. D. Wolf (1976). *Effects of Deicing Salts on Plant Biota and Soil; Experimental Phase.* Report 170, National Cooperative Highway Research Program. Washington, D.C.: Transportation Research Board.

Hanson, Andrew C. (1991). *Analysis of Energy Dissipation Caused by Snow Compaction during Displacement Plowing.* Transportation Research Record 1304, 177-81. Washington, D.C.: Transportation Research Board.

Hearst, Peter J. (1957). *Deicing Materials for Military Runways.* Tech. Memo. M-124. Port Hueneme, Calif.: U.S. Naval Civil Engineering Research and Evaluation Laboratory.

Hegmon, R. R., and W. E. Meyer (1968). *The Effectiveness of Antiskid Materials.* Highway Research Record No. 227. Washington, D.C.: Highway Research Board, pp. 50–56.

Hobbs, Peter V. (1974). *Ice Physics.* Oxford: Clarendon Press.

Hogbin, L. E. (1966). *Loss of Salt Due to Rainfall on Stockpiles Used for Winter Road Maintenance.* RRL Report No. 30. Harmondsworth, U.K.: Road Research Laboratory.

Howe-Grant, Mary (ed.) (1992). *Encyclopedia of Chemical Technology* (Kirk-Othmer). 4th ed. New York: John Wiley & Sons.

Hulme, M. (1982). A new winter index and geographical variations in winter weather. *Journal of Meteorology* 7(73), 294–300.

Hutchinson, Frederick E. (1972). Dispersal of soil-bound sodium from highway salting. *Public Works,* February, pp. 69–70.

Hutchinson, Frederick E. (1973). Discussion. In G. H. Brandt, *Potential Impact of Sodium Chloride and Calcium Chloride De-icing Mixtures on Roadside Soils and Plants.* Highway Research Record 425. Washington, D.C.: Highway Research Board, pp. 62–63.

ICAO (1995) *Aerodromes.* Annex 14 to the Convention on International Civil Aviation. Vol. 1, *Aerodrome design and operations.* Montreal, Que., Canada: International Civil Aviation Organization.

Ichihara, Kaoru, and Makoto Mizoguchi (1970). *Skid Resistance of Snow- or Ice-covered Roads.* Spec. Rept. 115. Washington, D.C.: Highway Research Board, pp. 104–114.

Ishimoto, Keishi (1995). *Studies on the Visibility Fluctuation by Airborne Snow Particles.* Report no. 107. Sapporo (Hokkaido), Japan: The Civil Engineering Research Institute, Hokkaido Development Bureau.

Ivanovskii, M. N., V. P. Sorokin, and I. V. Yagodkin (1982). *The Physical Principles of Heat Pipes.* Translated by R. Berman. New York: Oxford University Press.

Johnson, L. F. (1946). Ice prevention on New Hampshire highways. *Engineering News-Record,* April 4, pp. 86–88.

Ketcham, Stephen A., L. David Minsk, Robert R. Blackburn, and E. J. Fleege (1996). *Manual of Practice for an Effective Anti-icing Program: A Guide for Highway Winter Maintenance Personnel.* Report no. FHWA-RD-95-202. Washington, D.C.: Federal Highway Administration.

Kinsey, J. S., P. Englehart, M. A. Grelinger, K. Connery, C. Cowherd Jr., and J. Jones (1990). *Guidance Document for Selecting Antiskid Materials Applied to Ice- and Snow-covered Roadways.* EPA-450/3-90-007. Research Triangle Park, N.C.: U.S. Environmental Protection Agency. (Also issued with essentially the same information and corporate author Emission Standards Division as report no. EPA-450/3-90-007A, July 1991.)

Kinsey, John S. (1993). *Characterization of PM-10 Emissions from Antiskid Materials Applied to Ice- and Snow-covered Roadways.* EPA-600/R-93-019. Research Triangle Park, N.C.: U.S. Environmental Protection Agency.

Kinsey, John S. (1995). *Characterization of PM-10 Emissions from Antiskid Materials Applied to Ice- and Snow-covered Roadways—Phase II.* EPA-600/R-95-119. Research Triangle Park, N.C.: U.S. Environmental Protection Agency.

Koh, Gary (1989). *Physical and Optical Properties of Falling Snow.* Report 89-16. Hanover, N.H.: U.S. Army Cold Regions Research and Engineering Laboratory, p. 2.

Kuhajek, E. J., and H. W. Fiedelman (1976). *Behavior of Ferrocyanide and Cyanide in Relation to Deicing Salt Runoff.* Transportation Research Record 576. Washington, D.C.: Transportation Research Board, p. 33.

Kuroiwa, Daisuke (1958). *Icing and Snow Accretion.* Monograph Ser. No. 6. Sapporo, Japan: Hokkaido University, Res. Inst. of Appl. Electricity, pp. 1–30.

Lemon, Harold (1975). *1974–1975 Prewetted Salt Report.* Lansing: Michigan Department of State Highways and Transportation.

Lenga, Robert E. (1988). *The Sigma-Aldrich Library of Chemical Safety Data.* 2nd ed. Milwaukee, Wis: Sigma-Aldrich Corp. (P.O. Box 355, Milwaukee 53201).

Lewis, Russell M. (1983). *Practical Guidelines for Minimizing Tort Liability.* NCHRP Synthesis of Highway Practice 106. Washington, D.C.: Transportation Research Board.

Ludema, K. C., and B. D. Gujrati (1973). *An Analysis of the Literature on Tire-Road Skid Resistance.* SP-541. Philadelphia: American Society for Testing and Materials.

Ludlum, David M. (1982). *The American Weather Book,* vol. 2. Boston: Houghton Mifflin, p. 100.

Magono, C., and C. W. Lee (1966). Meteorological classification of natural snow crystals. *J. Faculty of Sci., Hokkaido University* 2, 321–335.

McCullough, David (1978). *The Path between the Seas: The Creation of the Panama Canal, 1870–1914.* New York: Simon & Schuster.

McFarland, Beverly L., and Kirk T. O'Reilly (1972). Environmental impact and toxicological characteristics of calcium magnesium acetate. In Frank M. D'Itri, ed., *Chemical Deicers and the Environment,* Boca Raton, Fla.: Lewis Publishers.

McKelvey, Blake (1995). *Snow in the Cities: A History of America's Urban Response.* Rochester, N.Y.: University of Rochester Press.

Merk (1989). *The Merck Index,* 11th ed. Rahway, N.J.: Merck & Co., Inc.

Minsk, L. D. (1970). *A Short History of Man's Attempts to Move through Snow.* Spec. Rept. 115. Washington, D.C.: Highway Research Board, pp. 1–7.

Mortimer, Thomas P., and Kenneth C. Ludema (1972). *The Effects of Salts on Road Drying Rates, Tire Friction, and Invisible Wetness.* Highway Research Record No. 396. Washington, D.C.: Highway Research Board, pp. 45–58.

Mudholkar, Vinay V. (1991). Preventing ice formation in tunnels. *Railway Track & Structures*, January, p. 26–27.
Nakaya, Ukichiro (1954). *Snow Crystals, Natural and Artificial.* Cambridge, Mass.: Harvard University Press, p. 114.
Nixon, Wilfrid A. (1993). *Improved Cutting Edges for Ice Removal.* SHRP-H-346. Washington, D.C.: Strategic Highway Research Program, National Research Council. (Available from Transportation Research Board, 2101 Constitution Ave., N.W., Washington, D.C. 20418.)
Pell, Kynric M. (1994). *An Improved Displacement Plow.* SHRP-H-673. Washington, D.C.: Strategic Highway Research Program, National Research Council. (Available from Transportation Research Board, 2101 Constitution Ave., N.W., Washington, D.C. 20418.)
Pfeifer, D. W., and M. J. Scali (1981). *Concrete Sealers for Protecting Bridge Structures.* NCHRP Report 244. Washington, D.C.: Transportation Research Board.
Pletan, R. A. (1972). *Salt Runoff Control at Stockpile Sites.* St. Paul: Minnesota Department of Transportation.
Powell, Kevin, Charles Reed, Lynne Lanning, and Dan Perko (1992). *The Use of Trees and Shrubs for Control of Blowing Snow in Select Locations along Wyoming Highways.* Report FHWA-92-WY-001. Cheyenne: Wyoming Highway Department.
Ringer, T. R. (1979). Protection methods for railway switches in snow conditions. In *Snow Removal and Ice Control Research,* Spec. Rept. 185. Washington, D.C.: Transportation Research Board, pp. 308–313.
SAE (1991). *Fluid, Deicing/Anti-Icing, Runways and Taxiways Potassium Acetate Base.* AMS 1432A. Warrendale, Penn.: Society of Automotive Engineers, Inc.
SAE (1992). *Compound, Solid Deicing/Anti-Icing Runways and Taxiways.* AMS 1431A. Warrendale, Penn.: Society of Automotive Engineers, Inc.
SAE (1993). *Fluid Deicing/Anti-Icing, Runways and Taxiways Glycol Base.* AMS 1426C. Warrendale, Penn.: Society of Automotive Engineers, Inc.
SAE (1994). *Sand, Airport Snow and Ice Control.* AMS 1448A. Warrendale, Penn.: Society of Automotive Engineers, Inc.
SAE (1995). *Fluid, Generic, Deicing/Anti-Icing Runways and Taxiways.* AMS 1435. Warrendale, Penn.: Society of Automotive Engineers, Inc.
Savas, E. S. *(1973).* The political properties of crystalline H2O: Planning for snow emergencies in New York. *Management Science* 20(2), 137–145.
Schaerer, P. A. (1970). *Compaction or Removal of Wet Snow by Traffic.* Spec. Rept. 115. Washington, D.C.: Highway Research Board, pp. 97–103.
Schneider, T. R. (1959). *Schneeverwehungen und Winterglätte (Snowdrifts and Winter Ice on Roads).* Interner Bericht Nr. 302. Eidgenößisches Institut für Schnee- und Lawinenforschung. Translated by D. A. Sinclair. Technical Translation TT-1038, National Research Council of Canada, Ottawa, July 1962.
Schneider, T. R. (1960). *Die Berechnung der zur Auflösung von Schnee- und Eiskrusten notwendigen Salztreumengen (The Calculation of the Amount of Salt Required to Melt Ice and Snow on Highways).* Interner Bericht Nr. 328. Eidgenößisches Institut für Schnee- und Lawinenforschung Weissfluhjoch-Davos. Translation TT-1004, National Research Council of Canada, 1962.
Shaw, Dale L. (1989). *Living Snow Fences: Protection that Just Keeps Growing.* Colorado Interagency Living Snow Fence Program. Fort Collins: Colorado State University.

Shoop, Sally A. (1993). *Terrain Characterization for Trafficability*. Report 93-6. Hanover, N.H.: U.S. Army Cold Regions Research and Engineering Laboratory.

Shoop, Sally A. (1993). *Three Approaches to Winter Traction Testing*. Report 93-9. Hanover, N.H.: U.S. Army Cold Regions Research and Engineering Laboratory.

Siple, Paul A., and Charles F. Passel (1945). Measurements of dry atmospheric cooling in subfreezing temperatures. *Proc. Am. Philosophical Soc. 89*(1), 177–199.

Stuart, K. D., and W. S. Mogawer (1991). *Laboratory Evaluation of Verglimit and PlusRide*. Report No. FHWA-RD-91-013. Washington, D.C.: Federal Highway Administration.

Sucoff, Edward (1975). *Effect of Deicing Salts on Woody Vegetation along Minnesota Roads*. Final Report, Investigation No. 636, University of Minnesota College of Forestry. St. Paul: Minnesota Department of Transportation.

Sullivan, J. L. (1831). Method of clearing the Baltimore rail-way of snow during the late winter. Am. J. Sci. 20, 166 (July). (no. 24 of Miscellanies).

Tabler, Ronald D. (1991). *Snow Fence Guide*. Report SHRP-H-320. Washington, D.C.: Strategic Highway Research Program, National Research Council. (Available from Transportation Research Board, 2101 Constitution Ave., N.W., Washington, D.C. 20418.)

Tabler, Ronald D. (1994). *Design Guidelines for the Control of Blowing and Drifting Snow*. Report SHRP-H-381. Washington, D.C.: Strategic Highway Research Program, National Research Council. (Available from Transportation Research Board, 2101 Constitution Ave., N.W., Washington, D.C. 20418.)

TRB (1991). *Highway Deicing: Comparing Salt and Calcium Magnesium Acetate*. Special Report 235. Washington, D.C.: Transportation Research Board.

Vershueren, Karel (1996). *Handbook of Environmental Data on Organic Chemicals*. 3rd ed. New York: Van Nostrand Reinhold.

Wallace, John M., and Peter V. Hobbs (1977). *Atmospheric Science. An Introductory Survey*. New York: Academic Press, p. 199.

Walvatne, Paul G. A. (1992). *A Case Study. Mn/DOT's Living Snow Fence Efforts*. Preprint, FHWA/SHRP Snow and Ice Control Seminar, Chicago, Sept. 28, 1992.

Welch, Bob H. (1976). *Economic Impact of Highway Snow and Ice Control—State of the Art*. Report No. FHWA-RD-77-20. Washington, D.C.: Federal Highway Administration.

Williams, G. P. (1976). *Design Heat Requirements for Embedded Snow-Melting Systems in Cold Climates*. Transportation Research Record 576. Washington, D.C.: Transportation Research Board, pp. 20–32.

Williamson, P. J. (1969). *The Formation and Treatment of Ice on Road Surfaces*. Paper 7224S, The Institution of Civil Engineers (UK).

Winterrowd, W. H. (1920). The development of snow fighting equipment. *Railway Maintenance Engineering 16*, 458–462.

References

Mudholkar, Vinay V. (1991). Preventing ice formation in tunnels. *Railway Track & Structures*, January, p. 26–27.

Nakaya, Ukichiro (1954). *Snow Crystals, Natural and Artificial.* Cambridge, Mass.: Harvard University Press, p. 114.

Nixon, Wilfrid A. (1993). *Improved Cutting Edges for Ice Removal.* SHRP-H-346. Washington, D.C.: Strategic Highway Research Program, National Research Council. (Available from Transportation Research Board, 2101 Constitution Ave., N.W., Washington, D.C. 20418.)

Pell, Kynric M. (1994). *An Improved Displacement Plow.* SHRP-H-673. Washington, D.C.: Strategic Highway Research Program, National Research Council. (Available from Transportation Research Board, 2101 Constitution Ave., N.W., Washington, D.C. 20418.)

Pfeifer, D. W., and M. J. Scali (1981). *Concrete Sealers for Protecting Bridge Structures.* NCHRP Report 244. Washington, D.C.: Transportation Research Board.

Pletan, R. A. (1972). *Salt Runoff Control at Stockpile Sites.* St. Paul: Minnesota Department of Transportation.

Powell, Kevin, Charles Reed, Lynne Lanning, and Dan Perko (1992). *The Use of Trees and Shrubs for Control of Blowing Snow in Select Locations along Wyoming Highways.* Report FHWA-92-WY-001. Cheyenne: Wyoming Highway Department.

Ringer, T. R. (1979). Protection methods for railway switches in snow conditions. In *Snow Removal and Ice Control Research*, Spec. Rept. 185. Washington, D.C.: Transportation Research Board, pp. 308–313.

SAE (1991). *Fluid, Deicing/Anti-Icing, Runways and Taxiways Potassium Acetate Base.* AMS 1432A. Warrendale, Penn.: Society of Automotive Engineers, Inc.

SAE (1992). *Compound, Solid Deicing/Anti-Icing Runways and Taxiways.* AMS 1431A. Warrendale, Penn.: Society of Automotive Engineers, Inc.

SAE (1993). *Fluid Deicing/Anti-Icing, Runways and Taxiways Glycol Base.* AMS 1426C. Warrendale, Penn.: Society of Automotive Engineers, Inc.

SAE (1994). *Sand, Airport Snow and Ice Control.* AMS 1448A. Warrendale, Penn.: Society of Automotive Engineers, Inc.

SAE (1995). *Fluid, Generic, Deicing/Anti-Icing Runways and Taxiways.* AMS 1435. Warrendale, Penn.: Society of Automotive Engineers, Inc.

Savas, E. S. *(1973).* The political properties of crystalline H2O: Planning for snow emergencies in New York. *Management Science* 20(2), 137–145.

Schaerer, P. A. (1970). *Compaction or Removal of Wet Snow by Traffic.* Spec. Rept. 115. Washington, D.C.: Highway Research Board, pp. 97–103.

Schneider, T. R. (1959). *Schneeverwehungen und Winterglätte (Snowdrifts and Winter Ice on Roads).* Interner Bericht Nr. 302. Eidgenößisches Institut für Schnee- und Lawinenforschung. Translated by D. A. Sinclair. Technical Translation TT-1038, National Research Council of Canada, Ottawa, July 1962.

Schneider, T. R. (1960). *Die Berechnung der zur Auflösung von Schnee- und Eiskrusten notwendigen Salztreumengen (The Calculation of the Amount of Salt Required to Melt Ice and Snow on Highways).* Interner Bericht Nr. 328. Eidgenößisches Institut für Schnee- und Lawinenforschung Weissfluhjoch-Davos. Translation TT-1004, National Research Council of Canada, 1962.

Shaw, Dale L. (1989). *Living Snow Fences: Protection that Just Keeps Growing.* Colorado Interagency Living Snow Fence Program. Fort Collins: Colorado State University.

Shoop, Sally A. (1993). *Terrain Characterization for Trafficability.* Report 93-6. Hanover, N.H.: U.S. Army Cold Regions Research and Engineering Laboratory.

Shoop, Sally A. (1993). *Three Approaches to Winter Traction Testing.* Report 93-9. Hanover, N.H.: U.S. Army Cold Regions Research and Engineering Laboratory.

Siple, Paul A., and Charles F. Passel (1945). Measurements of dry atmospheric cooling in subfreezing temperatures. *Proc. Am. Philosophical Soc.* 89(1), 177–199.

Stuart, K. D., and W. S. Mogawer (1991). *Laboratory Evaluation of Verglimit and PlusRide.* Report No. FHWA-RD-91-013. Washington, D.C.: Federal Highway Administration.

Sucoff, Edward (1975). *Effect of Deicing Salts on Woody Vegetation along Minnesota Roads.* Final Report, Investigation No. 636, University of Minnesota College of Forestry. St. Paul: Minnesota Department of Transportation.

Sullivan, J. L. (1831). Method of clearing the Baltimore rail-way of snow during the late winter. Am. J. Sci. 20, 166 (July). (no. 24 of Miscellanies).

Tabler, Ronald D. (1991). *Snow Fence Guide.* Report SHRP-H-320. Washington, D.C.: Strategic Highway Research Program, National Research Council. (Available from Transportation Research Board, 2101 Constitution Ave., N.W., Washington, D.C. 20418.)

Tabler, Ronald D. (1994). *Design Guidelines for the Control of Blowing and Drifting Snow.* Report SHRP-H-381. Washington, D.C.: Strategic Highway Research Program, National Research Council. (Available from Transportation Research Board, 2101 Constitution Ave., N.W., Washington, D.C. 20418.)

TRB (1991). *Highway Deicing: Comparing Salt and Calcium Magnesium Acetate.* Special Report 235. Washington, D.C.: Transportation Research Board.

Vershueren, Karel (1996). *Handbook of Environmental Data on Organic Chemicals.* 3rd ed. New York: Van Nostrand Reinhold.

Wallace, John M., and Peter V. Hobbs (1977). *Atmospheric Science. An Introductory Survey.* New York: Academic Press, p. 199.

Walvatne, Paul G. A. (1992). *A Case Study. Mn/DOT's Living Snow Fence Efforts.* Preprint, FHWA/SHRP Snow and Ice Control Seminar, Chicago, Sept. 28, 1992.

Welch, Bob H. (1976). *Economic Impact of Highway Snow and Ice Control—State of the Art.* Report No. FHWA-RD-77-20. Washington, D.C.: Federal Highway Administration.

Williams, G. P. (1976). *Design Heat Requirements for Embedded Snow-Melting Systems in Cold Climates.* Transportation Research Record 576. Washington, D.C.: Transportation Research Board, pp. 20–32.

Williamson, P. J. (1969). *The Formation and Treatment of Ice on Road Surfaces.* Paper 7224S, The Institution of Civil Engineers (UK).

Winterrowd, W. H. (1920). The development of snow fighting equipment. *Railway Maintenance Engineering* 16, 458–462.

Glossary of Frequently Used Terms

adhesion (pave) (chem)* The attraction between dissimilar materials, e.g., the bonding that occurs between ice and pavement.

anti-icing: (pave) Snow and ice control technique for preventing the formation of a strong bond between ice or compacted snow and the pavement on which it lies, either by use of a chemical freezing-point depressant or by thermally maintaining the pavement at or above the freezing point.

Average Daily Traffic (ADT): The average number of vehicles passing a fixed point in a 24-h period; it is a convention for measuring traffic volume.

axle ratio: (truck) The number of revolutions made by the input driving gear for each revolution of the output driven gear; also known as *gear ratio*.

benching: Process of reducing the height of roadside windrows by pushing back the upper part, leaving a bench or shelf to increase storage capacity for a following storm; also called *shelving* and *high winging*.

BOD (biochemical oxygen demand): The quantity of oxygen (in milligrams per liter) consumed during the biodegradation of organic matter over a specified period of time, usually the first five days. A high

---

*Entries referring to a usage by industry or profession are prefaced by:
    chem   chemistry
    met   meteorology
    pave   pavement condition
    truck   truck industry

level is usually associated with a low availability of dissolved oxygen, and is detrimental to aquatic life.

black ice: (met)(pave) Popular term for a very thin coating of clear, bubble-free, homogeneous ice which forms on a pavement with a temperature at or slightly above 32°F (0°C) when the temperature of the air in contact with the ground is below the freezing point of water and small, slightly supercooled water droplets are deposited on the surface and coalesce (flow together) before freezing.

blizzard: (met) Snow falling at the rate of 1 in/h (2.5 cm/h) or more with wind speed of at least 35 mi/h (15.6 m/s) (NWS).* According to an item in *American Speech* in 1928, the first appearance of this word referring to a snow squall occurred in 1870 in the Estherville [Iowa] *Vindicator*.

blowing snow: (met) Old snow picked up by the wind from already deposited accumulations and transported across a road. Sometimes called a *ground blizzard*.

bogie: (truck) A tandem rear-axle suspension assembly.

bolted frame assembly: (truck) A frame assembled with nuts, bolts, and washers rather than rivets or welds.

braking torque: (truck) The torque exerted on the axle housing during braking and transferred to the frame by the rear axle's suspension or torque arms.

brine: A solution of a salt (the solute) and water (the solvent), generally saturated or at a high concentration.

change of state: (chem) The transformation from one form to another; water substance may undergo six transformations, each of which is identified by a specific term: (1) condensation, the change from water vapor to liquid, (2) evaporation, the change from liquid to vapor, (3) freezing, the change from liquid to solid, (4) melting, the change from solid to liquid, (5) sublimation, the change from solid directly to vapor, and (6) deposition, the change directly from vapor to solid.

COD (chemical oxygen demand): Measurement of all the oxidizable matter found in a runoff sample, a portion of which could deplete dissolved oxygen in receiving waters.

coefficient of friction: *See* friction.

cohesion: (chem) The attraction between similar materials, e.g., the bonding that occurs within a snowpack.

---

*Sources for entries are indicated where appropriate by:
    ASTM   American Society for Testing and Materials
    EPA   U.S. Environmental Protection Agency
    NWS   U.S. National Weather Service

crossmember: (truck) A structural metal shape that ties together the side rails of a truck frame. Crossmembers hold the side rails in parallel alignment and resist twisting. A crossmember may support or help to support other components, such as the engine, transmission, fuel tanks, and battery boxes.

curb weight: (truck) Weight of a fully equipped vehicle without any payload; also known as *tare weight*. The difference between a vehicle's curb weight and GVWR is its maximum allowable payload weight.

damp: (pave) Light coating of moisture on the pavement, resulting in slight darkening of PCC, but with no visible water drops.

degree-day: (met) A measure of the departure of the mean daily temperature from a given standard. For heating and cooling, the base is 65°F, with one degree-day being recorded for each degree above (for cooling) or below (for heating). Degree-days are accumulated over a "season," at any point during which the total can be used as an index of past temperature relations with some quantity, such as plant growth, fuel consumption, power output, etc. (NWS).

deicing: (pave) Snow and ice control technique for destroying the bond between ice or compacted snow and the pavement as the initial step in removal.

dew: (met) Water vapor condensed onto grass and other objects near the ground (NWS).

dew point: (met) The temperature at which water vapor will condense on an object. Dew point is used as a measure of moisture in the air. Higher dew points indicate more moisture. Dew points above 60°F are uncomfortable for most people (NWS).

differential lockout: (truck) Gearing in a rear axle that transmits power to both rear wheels, yet allows one wheel to turn more slowly than the other when turning a corner; also called *power divider lockout*.

drawbar pull: (truck) Tractive effort in excess of that required for forward motion.

driveline: (truck) Torque-transmitting components linking the transmission and rear axle, including flanges, end yokes, U-joints, slip joints, and propeller shaft.

drizzle: (met) Very small water droplets with diameter between 0.008 and 0.04 in (200 and 500 μm). Note: The diameter of a human hair is about 0.003 in (70 μm).

dry: (pave) No wetting of the pavement surface.

echelon plowing: A group of snowplows traveling down a highway in stepwise formation, each in a different path, to move the snow to either side or both sides of the road depending on number of lanes, storage capacity of the roadsides, and pavement drainage.

endothermic: Absorbing heat during a chemical reaction.

exothermic: Releasing heat during a chemical reaction.

fine aggregate: Aggregate passing the $\frac{3}{8}$-in sieve and almost entirely passing the No. 4 (4.75-mm) sieve while predominately retained on the No. 200 (75-μm) sieve (ASTM).

flange: (truck) The horizontal top or bottom portion of a frame siderail.

frame: (truck) Generally a pair of C-channel members joined with steel crossmembers in a ladder shape, normally 34 in (356 mm) wide, which supports the engine, driveline, cab, body, suspension components, and payload.

frame cutoff, maximum: (truck) The shortest allowable extension behind the rear axle, established for each wheelbase by manufacturers and printed in their data books and in body builders' books.

fog: (met) A cloud touching the Earth's surface. When made up of ice crystals, it is called an ice fog (NWS).

freezing rain: (met) Supercooled droplets of liquid precipitation falling on a surface whose temperature is below or slightly above freezing, resulting in a hard, slick, generally thick coating of ice commonly called *glaze* or *clear ice*. Nonsupercooled raindrops falling on a surface whose temperature is well below freezing will also result in glaze.

friction: (*a*) Friction force, the resisting force parallel to the sliding surface and at the interface between two bodies that is produced when, under the action of an external force, one body moves relative to the other. (*b*) Coefficient of friction, the ratio of the force resisting the sliding motion between two bodies to the force pressing the two bodies together. It is usually denoted by the Greek letter μ (mu). Values range from 0 to 1.0 (*see* skid number).

frost: (met) Ice crystals in the form of scales, needles, feathers, or fans deposited on surfaces cooled by radiation or by other processes. The deposit may be composed of drops of dew frozen after deposition and of ice formed directly from water vapor at a temperature below 32°F (0°C). Also called *hoarfrost*.

glaze ice: (met) A coating of ice thicker than so-called black ice which is formed from *freezing rain*, or from freezing of ponded water or poorly drained meltwater. It may be clear or milky in appearance, but is denser, harder and more transparent than either rime or hoarfrost, and generally is smooth, though it sometimes may be rough.

GAWR: (truck) Gross axle weight rating; the rated capacity of the axle, suspension, tires, and wheels.

gradability: (truck) The maximum grade a vehicle can climb at a given speed and GVW.

GVW: (truck) Gross vehicle weight; the total weight of the fully equipped and loaded vehicle, including weight of truck and payload, engine fluids, fuel, and driver.

GVWR: (truck) Gross vehicle weight rating; the maximum gross vehicle weight (GVW) the vehicle is designed to handle.

heat of fusion: *See* latent heat of fusion.

heat of vaporization: *See* latent heat of vaporization.

heat pipe: A closed device in which the latent heat of vaporization is transferred by evaporating a working fluid in the heat input region and condensing the vapor in the heat discharge region; the liquid is returned to the heat input area by capillary action in a wick structure.

heat pump: An assembly of a refrigerant gas compressor, heat exchangers, piping, controls, and accessories which can provide heating or cooling; the basic types are air-to-air, water-to-air, water-to-water, earth-to-air, and earth-to-water.

hoarfrost: (met) Ice crystals formed on objects exposed to the air, such as tree branches, plant stems, wires, poles, etc. It is the result of water vapor passing from a gas to a solid form (NWS). It commonly appears as a white deposit of ice crystals formed by the freezing of dew. Also called *white frost. Hoar* comes from an Old English word meaning "gray" or "grayish-white" and was used to describe this deposit as early as the thirteenth century.

horsepower: (truck) The work an engine can do in a specific time, equal to the work of lifting 33,000 pounds a distance of one foot in one minute, or 33,000 ft·lb/min; it depends on engine torque and revolutions per minute (rpm). Brake horsepower is the actual horsepower delivered by the crankshaft, calculated as:

$$\text{Brake horsepower} = (\text{torque} \times \text{rpm})/5252$$

hygroscopic: For a chemical, having a tendency to attract water.

ice control chemical: A chemical used for either anti-icing or deicing.

interaxle differential: (truck) A differential installed between the front and rear axles of a tandem assembly that allows for equal power to each axle; also called *power divider*.

latent heat of fusion: The heat required to melt a unit mass of pure ice, 144 Btu/lb or 334 J/g.

latent heat of vaporization: The heat required to evaporate a unit mass of water to water vapor, 1078 Btu/lb or 2500 J/g.

LD$_{50}$: The amount of a chemical that is lethal to one-half (50 percent) of the experimental animals exposed to it. LD$_{50}$ is usually expressed as the weight of the chemical per unit of body weight (mg/kg). It may be fed (oral LD$_{50}$), applied to the skin (dermal LD$_{50}$), or administered in the form of vapors (inhalation LD$_{50}$).

light rain: (met) Liquid droplets small in size falling at a rate insufficient to result in standing water (puddling) or visible runoff from a road.

light snow: (met) Snow falling at the rate of less than ½ in (12 mm) per hour; visibility is not affected adversely.

loose snow: (met) Unconsolidated snow, i.e., snow lacking intergranular bonds which can be easily blown into drifts or off of a surface.

manufactured sand: The fine material resulting from the crushing and classification by screening, or otherwise, of rock, gravel, or blast furnace slag (ASTM).

M.I.C.: Mineral insulated cable used for resistance heating; also called *MI cable*.

NWS: National Weather Service, an agency of the U.S. National Oceanic and Atmospheric Administration.

packed snow: The infamous "snowpack" or "pack" which results from compaction of wet snow by traffic or by alternate surface melting and refreezing of the water which percolated through the snow or which flowed from poorly drained shoulders.

particulates: (1) Fine liquid or solid particles, such as dust, smoke, mist, fumes, or smog, found in air or emissions. (2) Very small solids suspended in water. They vary in size, shape, density, and electric charge, and can be gathered together by coagulation and flocculation (EPA).

permeability: The rate at which liquids pass through soil or other materials in a specified direction (EPA).

pH: An expression of the intensity of the basic or acid condition of a liquid. The pH may range from 0 to 14, where 0 is the most acid and 7 is neutral. Natural waters usually have a pH between 6.5 and 8.5 (EPA).

phase diagram: (chem) A chart that describes the equilibrium states of matter. It relates the pressure and temperature of a substance such as water. Solubility diagrams which describe the solubility and freezing point of a solution as a function of the amount of solute (Figs. 3.3 and 3.4) are also popularly called phase diagrams.

PM$_{10}$: The standard for measuring the amount of solid or liquid matter suspended in the atmosphere, i.e., the amount of particulate matter over 10 μm (micrometers) (0.0000394 in) in diameter; smaller PM$_{10}$ particles penetrate to the deeper portions of the lung, affecting sensitive population groups such as children and individuals with respiratory ailments (EPA).

policy manual: A document describing an agency's practices for responding to winter maintenance requirements, expressed in general terms for the public or for administrative purposes, with specific details of implementation left for operating procedure manuals.

pollutant: Generally, any substance introduced into the environment that adversely affects the usefulness of a resource (EPA).

pollution: Generally, the presence of matter or energy whose nature, location, or quantity produces undesired environmental effects. Under the Clean Water Act, for example, the term is defined as the manmade or man-induced alteration of the physical, biological, chemical, and radiological integrity of water (EPA).

porosity: Degree to which soil, gravel, sediment, or rock is permeated with pores or cavities through which water or air can move (EPA).

Prevention of Significant Deterioration (PSD): EPA program in which state and/or federal permits are required in order to restrict emissions from new or modified sources in places where air quality already meets or exceeds primary and secondary ambient air quality standards (EPA).

primary standards: National ambient air quality standards designed to protect human health with "an adequate margin for safety" (EPA).

probability of precipitation (POP): The likelihood of occurrence (in percent) of a precipitation event at any given point in the forecast area. Two different methods are used to indicate the chance of precipitation for a specific area: numerical or nonnumerical terms. The "expression of uncertainty" category is used for widespread precipitation, and the "equivalent areal coverage" is used for convective (i.e., showery) events. Below is a table of these two methods with the corresponding POP (NWS).

POP, %	Expression of uncertainty	Equivalent areal coverage
0	None used	None used
10	Slight chance (seldom used)	Isolated or few
20	Slight chance	Widely scattered
30–50	Chance	Scattered
60–70	Likely	Numerous
80–100	None used	None used

Other qualifying terms used with these nonnumerical expressions:

Duration: brief, occasional, intermittent, frequent.

Intensity:

Very light	Less than 0.01 in (0.25 mm)
Light	0.01 to 0.10 inch (0.25–2.5 mm) per hour
Moderate	0.10 to 0.30 inch (2.5–7.6 mm) per hour
Heavy	Greater than 0.30 inch (7.6 mm) per hour

procedure manual: A document detailing an agency's practices and procedures regarding personnel, equipment, and materials used for winter maintenance, including prestorm preparations, mobilization, operations, and cleanup.

rain: (met) Liquid precipitation falling at a rate sufficient to result in noticeable flow from a road surface or along a road gutter.

RBM: (truck) Resisting bending moment, a measure of the strength of truck frame side rails, calculated as the product of yield strength and section modulus; the higher the RBM, the stronger the member.

relative humidity: (met) The ratio of the air's water vapor content to the maximum water vapor it could hold at a given temperature (NWS).

rime: (met) A white granular deposit of ice on the windward sides of objects, always facing directly into the wind. It is denser and harder than hoarfrost, but lighter, softer, and less transparent than glaze (NWS). It is formed by supercooled water droplets of fog contacting a solid object at a temperature below 32°F (0°C).

road weather information system (RWIS): An aggregation of sensing and processing software and equipment, communications, and forecasting support, generally consisting of on-site weather and pavement condition sensors and off-site meteorological processing, used for providing timely and road-specific weather forecasts.

section modulus: (truck) A measure of the strength of frame side rails determined by cross-sectional area and shape.

skid number: The coefficient of friction μ multiplied by 100.

sleet: (met) There is no international agreement on this term; in American usage it generally refers to ice pellets, whereas in European usage it is used to describe a mixture of liquid and frozen precipitation or of snow partially melting as it falls.

slurry: A mixture of water and finely divided insoluble material.

slush: (pave) Accumulation of snow which lies on an impervious base and is saturated with water in excess of its freely drained capacity. It will not support any weight when stepped or driven on but will "squish" until the base support is reached.

Glossary

snow: (met) Snow falling at a rate greater than ½ in/h (13 mm/h); visibility may be reduced.

snowburst or thundersnow: (met) An intense convective snow squall accompanied by lightning and thunder. Snowfall rates in snowbursts often reach 1 to 3 in/h (25–76 mm/h)(NWS).

startability: (truck) A vehicle's grade-pulling capability from a dead stop at full GVW.

sublimation: (met) The process whereby snow and ice pass directly to the vapor stage without first melting. Much of our snow and ice disappears this way in the winter (NWS).

tandem rear axle: (truck) A pair of drive axles linked by an interaxle propeller shaft.

temperature: A measure of the molecular energy of a substance (heat).

thermal mapping: The practice of recording pavement surface temperature by remote infrared scanning over a range of weather and temperature conditions and preparing plots of temperature related to geographical position; used for identifying problem locations such as cold sinks and for siting ice detectors.

tonne: A metric ton, 1000 kg, equal to 2204.6 lb or about 1.1 tons.

torque: (truck) The amount of twisting effort on the crankshaft, expressed as pound-feet (lb•ft), representing a force of one pound acting at a right angle on the end of a one-foot-long lever; torque is greatest at medium speeds but drops off as maximum horsepower is reached as a result of the rate at which fuel can enter the cylinders.

traction, net (truck)**:** The maximum force developed by the driving element minus the motion resistance; it is equal to drawbar pull.

wet: (pave) Road surface saturated with water from rain or meltwater, whether or not resulting in puddling or runoff.

wind chill: (met) A calculation of how cold it feels outside when the effects of temperature and wind speed are combined. A strong wind combined with a temperature of just below freezing can have the same effect as a still air temperature about 35°F (19.5°C colder (NWS). (See the chart in Appendix A.)

yield strength: (truck) The maximum stress to which a frame may be subjected through loading and still return to its original shape with no deformation upon removal of the stress.

Index

Abrasives:
  early use of, 42
  for friction, 85–89
  for ice control, 8
Absorption:
  in chemical tables, 246
  of materials, 4
Accidents:
  from snow and ice conditions, 143–144
  from snow removal, 111
Acetates:
  calcium magnesium acetate, 241, 248
  corrosion from, 58
  environmental effects of, 9
  potassium acetate, 243, 253–254
  sodium acetate, 243, 256–257
Active sensor devices, 119–120
Additives for sodium chloride, 221
Adhesiveness:
  of ice, 38–39
  of snow, 26–27
Age hardening, 3, 27–28
Agency resources:
  in decision making, 129
  interagency cooperation, 140–142
  for weather/climate information, 117
Agitation of snow, 28
Agreements, interagency, 141
Air blowers, 60
Air brakes, 166
Air quality, abrasives effects on, 89
Air temperature in decision making, 124
Aircraft deicing fluids, 195
Airport snow and ice control, 191–192
  application rates for, 204
  blade plows for, 192–193
  brooms for, 193
  chemicals for, 194–197
  costs of, 12
  friction testing and reporting, 197–199

Airport snow and ice control (*Cont.*):
  heating pavements in, 62
  liquid applicators for, 194
  reference materials for, 199–200
  rotary plows for, 193
  snow desks for, 199
  spreaders for, 193–194
All-wheel drive vehicles, 162
AMA 1431A specification, 242
Ammonium compounds, 9
Amontons, Guillaume, 79
Amundson, William M., 134
Angle of repose for salt, 56
Anti-icing:
  for airports, 195
  vs. deicing, 7–8
  for roadways, 91–93
Antifreezes, 49
Application methods and rates:
  of antiskid material, 88
  of chemicals, 51–53
    for airports, 195–197
    conversion factors for, 203–204
Applied loads in friction, 79
Aqueous solutions, 49
Ashes for friction, 86
Asphalt concrete, 9
Atmospheric sensors, 120
Augers in spreaders, 105
Automatic transmissions, 166
Auxiliary lighting systems for trucks, 168
Availability:
  in chemical tables, 49–50
  of materials, 5
Axles for snowplow trucks, 160–163

Bare road policies, 86
Beilhack plows, 189
Belden Tunnel, 187

279

Black ice:
  on bridge decks, 62
  formation of, 37
Blade plows, 192–193
Blades for snow removal vs. ice removal, 61–62
Block heaters, 165
Blowers:
  air, 60
  for railroad switches, 185–186
  snow, 172–173
Blowover, 61, 111
Bodies of snowplow trucks, 169–170
Bonding, 6
  in age hardening, 27–28
  hydrogen, 31
Bonding inhibitors, 8
Brakes, 166–167
Bridge decks, 62
Brine containment, 57
Brooms, 60
  for airports, 104, 193
  power, 61, 104
  speed of, 61
Brushes, 60
Bucker plows, 182–183
Buildings for salt storage, 54–57
Bulldozers, 185, 189

Cab-over-engine design, 167
Cables in heated pavements, 66
Cabs for snowplow trucks, 167–168
Calcium chloride, 209–211
  characteristics of, 247
  corrosion from, 58
  forms of, 211–212
  and friction, 85
  in pavements, 59
  phase diagram for, 47–48
  preparation of, 212–218
  for prewetting chemicals, 52
  in sand mixtures, 89
  sources for, 212
  standards for, 212
Calcium magnesium acetate, 241, 248
Calibration:
  of spreaders, 108
  of thermometers, 124
Canadian Hardness Gage, 23
Carriers, 174
Cast spoke wheels, 163

Celsius to Fahrenheit conversion table, 205–206
Central processing units in RWIS systems, 119
Centralization of responsibility, 134
Chatter, 110
Chemical Abstracts Service (CAS) Registry Number, 245
Chemical concentration detectors in RWIS systems, 119
Chemicals and chemical control methods, 41–44
  for airports, 194–197
  application methods for, 51–52
  application rates of, 53
  conversion factors for, 204
  corrosion from, 57–58
  in deicing vs. anti-icing, 8
  environmental effects of, 9
  friction improvements from, 90–93
  friction losses from, 85
  for ice control, 8, 244–261
  in ice melting process, 44–48
  in pavements, 59–60
  preparation of, 212–218
  prewetting, 52–53
  problem avoidance, 60
  selection of, 48–51
  spreaders for, 109
  storage and handling of, 53–57
  vegetation damage from, 58–59
  weight percent calculations for, 201
Chlorides:
  calcium chloride (*see* Calcium chloride)
  environmental effects of, 9
  magnesium chloride (*see* Magnesium chloride)
  sodium chloride (*see* Salt; Sodium chloride)
Cinders, 42, 87
Cities, snow disposal in, 149
Cleanup of antiskid material, 89
Clear-sight windows, 112
Climate factors:
  in chemical selection, 50
  in material selection, 5
Clinkers, 87
Closed-loop controllers for spreader spinners, 107–108
Coefficient of friction, 78–81
  for airports, 198
  for ice and snow, 42

Index

Cohesion of snow, 22, 26
Command and communication centers in decision making, 128–131
Communications:
  with public, 129–130, 147
  in RWIS systems, 118–121
Compaction, snow, 2–3, 23–25
Composition of materials, 4
Compressibility of snow, 23–25
Computers in RWIS systems, 119
Concrete:
  chemical effects on, 9
  for friction, 87
Condensation nuclei, 18
Conductive heated pavements, 66–67
Conductivity:
  conversion factors for, 202
  of snow, 29–30
  in thermal methods, 63
Cone spray nozzles, 52
Coniferous trees for snow fences, 74
Constructive notice, 138
Contact area in friction, 79
Continuous-friction measuring equipment, 197
Controllers for spreader spinners, 106–108
Convection in thermal methods, 63
Conversion factors, 201–206
Corn stubble as snow fences, 76
Corrosion current, 57
Corrosion effects:
  in chemical tables, 246
  of chemicals, 57–58
  of materials, 4
Corrosion inhibitors, 50
Cost:
  of airport delays, 191
  of chemicals, 49–50, 246
  of ice removal, 33–34
  of living snow fences, 75–76
  of materials, 5
  of snow and ice control, 11–12
  of truck attachments, 171–172
Cost/benefit analyses for trucks, 154
Covered bridges, 12
Covered storage for salt, 54–57
Crystal formation, 18–21
Cycle time of service vehicles, 142

D 98 calcium chloride standard, 212
D 632 sodium chloride standard, 219

da Vinci, Leonardo, 79
Dead Sea, magnesium chloride from, 240
Dead spinners in chemical application, 51
Deceleration in friction measurements, 83
Decelerometers at airports, 197–199
Decision making, 123–124
  command and communication centers for, 128–130
  command of operations in, 130–131
  criteria in, 124–127
  pavement friction status in, 127–128
  reporting requirements in, 131–132
Deformable moldboards, 100
Deicing:
  for airports, 195
  vs. anti-icing, 7–8
  for roadways, 93
Density:
  conversion factors for, 202
  of ice, 34
  of snow, 21–22
  and snow hardness, 22–23
Depth hoar, 27–28
Design factors:
  in snow fences, 71–72
  in thermal methods, 68–70
Design Guidelines for the Control of Blowing and Drifting Snow, 71
Detectors:
  for pavement temperature, 6–7
  in RWIS systems, 119–120
Dew, 36
Dew point, 36
Dew-point temperature sensors, 120
Direct radiation in thermal methods, 67–68
Disaggregation of snow, 28
Disaggregators in rotary plows, 102
Disc wheels, 163
Displacement plows, 2, 97–102
Disposal of snow, 70, 149
Distributors for spreaders, 106–108
Dolce, John E., *Fleet Manager's Guide to Specification and Procurement,* 156
Dome salt storage structures, 55–56
Doppler radar, 93, 116
Double-track railroad plows, 188
Drift control:
  benefits of, 70–71
  snow fences for, 71–76
Drivelines for snowplow trucks, 165–166
Dump boxes, 106

E 449 calcium chloride standard, 212
E 534 sodium chloride standard, 219
Electrically heated pavements, 66
Electrified railroad systems, 190
Elliot, J. W., 183
Embedded detectors, 6–7
Emergency Management Office, 141–142
Employee status:
  as decision making factor, 131
  reporting of, 147
Endothermic solids, 44
Energy:
  in compaction, 2–3, 24–25
  in ice melting process, 44
Engines for snowplow trucks, 164–165
Environmental effects:
  in chemical tables, 246
  of chemicals, 9
  corrosion, 57–58
  of ethylene glycol, 49
  as material selection criterion, 5
  of sodium chloride, 222
  of urea, 242
  on vegetation, 58–59
Environmental factors in friction, 79
Epoxy-coated rebars, 9
Equilibrium air water vapor, 35
Equipment:
  as decision-making factor, 131
  ice removal (see Ice and ice control)
  snow removal (see Snow removal equipment)
  vehicles (see Trucks; Vehicles)
Ethylene glycol:
  for airports, 195
  characteristics of, 242, 248–250
  environmental damage from, 49
Eutectic temperature, 4
  in chemical tables, 245
  of salt, 9, 47
Eutrophication, 59
Evaluation of performance, 149–150
Evaporation:
  in brine containment, 57
  in freezing process, 35
Exothermic solids, 44

Fahrenheit to Celsius conversion table, 205–206
Fan nozzles, 52
Federal-Aid Road Act, 97

Fences (see Snow fences)
Fetch in snow fence design, 72
Finch, James William, *Motor Truck Engineering Handbook*, 156
Finger drifts, 70–71
Fixed installations for railroad switches, 185–186
Fixed lateral friction, 84
Fixed-slip longitudinal braking friction, 84
Flangers for railroads, 184, 188
Flash icing, 37
Fleet Manager's Guide to Specification and Procurement (Dolce), 156
Form of materials, 4
Formates:
  corrosion from, 58
  environmental effects of, 9
  sodium formate, 243–244, 258–259
Formation mechanisms for ice, 34–37
Forrest, Nathan Bedford, 115
Frames for snowplow trucks, 156–160
Freeze-thaw of compacted snow, ice from, 35
Freezing nuclei, 18
Freezing point:
  in ice melting process, 44–47
  and pressure, 201
Freezing rain, 35
Friction:
  abrasives for, 85–89
  at airports, 197–199
  chemical methods for, 90–93
  coefficient of friction, 78–81
  critical values in, 84–85
  as decision-making factor, 127–128
  factors in, 79
  measurements of, 81–84
  mechanical methods for, 90–91
  in roadway maintenance, 77
  slip ratio in, 80–81
  thermal methods for, 93
Front-end loaders:
  for railroads, 185
  uses of, 112, 174
Front plows, 97–98
Frost formation, 36–37
Fuel tank warmers, 165
Fusion, latent heat of, 47–48

Gaugler, R. S., 65n
Geographical tracking systems, 148

Geothermal sources for heated pavements, 66
Glare ice, 35
Glaze ice, 35
Glossary of terms, 269–277
Glycol-based chemicals:
  for airports, 194–195
  environmental effects of, 49
Good Roads Movement, 96
Government weather/climate information sources, 116
Gradalls for railroads, 189
Graders:
  for ice control, 110
  uses of, 112, 173–174
Great Salt Lake, magnesium chloride from, 240
Gross weight of snowplow truck engines, 164
Grover, G. M., 65n

Halite, 218
Handling of chemicals, 53–57
Hardness of snow, 3, 22–24, 27–28
Heat capacity of snow, 28–29
Heat island effect, 12
Heat of solution for calcium chloride, 211
Heat pipes, 64–67
Heated pavements, 63–67
Heated wiper blades, 111
Heaters and blowers for railroad switches, 185–186
Heating (*see* Thermal methods)
Highway grade railroad crossings, 187
Hitches, 170–171
Hoarfrost formation, 36–37
Homogeneous nucleation, 36
Hoosac Tunnel, 187
Hoppers:
  for salt storage, 56
  for spreaders, 106
Horsepower of snowplow truck engines, 164
Human effects in chemical tables, 246
Humidity and sodium chloride, 221
Hydrates, 209
Hydraulic brake systems, 166–167
Hydraulic power systems, 169
Hydrogen bonding, 31
Hydronic heated pavement systems, 64–65
Hydroplaning, 95

Ice and ice control, 104
  adhesion of, 38–39
  characteristics of, 11, 33–34
  chemical processes in, 44–48
  chemicals for, 244–261
  costs of, 11–12
  formation mechanisms for, 34–37
  hoarfrost, 36–37
  ice-cutting blades for, 110
  mechanical methods for, 34, 61–62
  molecular structure of, 31–32
  physical properties of, 34
  principles of, 7–9
  in railroad tunnels, 187
  rime ice, 37–38
  spreaders for, 104–109
  underbody blades for, 110
Ice-cutting blades, 110
Ice detectors:
  for pavement temperature, 6–7
  in RWIS systems, 119–120
Impellers in rotary plows, 102–103
Infrared radiometers, 6, 121
Instability of snow, 27
Insulators:
  ice as, 34
  snow as, 28–29
Intelligence (*see* Weather/climate information)
Intensity of events as decision-making factor, 125–126
Interagency cooperation, 140–142
Ions in corrosion, 57–58
Irish setter friction test, 83
Isopropyl alcohol:
  characteristics of, 250–251
  uses of, 49

James Brake Decelerometers, 197
James Brake Index (JBI), 197
Jordan spreaders, 189
Jull, Orange, 184
Jungle telegraph, 146

Kingpost mounts, 101

Latent heat of fusion, 47–48
Legal considerations, 136–138
Leslie, Edward, 184

Leslie, John S., 184
Leslie plows, 184, 188–189
Level of service, 142–143
Liability, tort, 136–137
Lighting systems for snowplow trucks, 168–169
Liquid applicators for airports, 194
Liquid chemicals:
  application of, 51–52
  conversion factors for, 204
  for ice control, 8
  preparation of, 212–218
  selection of, 49
  spreaders for, 109
Liquidus curve of salt, 47
Living snow fences, 4, 74–76
Loading considerations for snowplow trucks, 158–159, 162–163
Locked wheel longitudinal braking friction, 84
Low-energy surfaces, 27
Lowest effective temperature:
  in chemical tables, 245
  of materials, 4

Magnesium chloride:
  characteristics of, 222, 240, 251–252
  corrosion from, 58
  for prewetting chemicals, 52
Main railroad lines, 184–185
Maintainability of trucks, 173
Management control, 134
Manual of Practice for an Effective Anti-icing Program, 93
Manuals, 134–136
Manufactured sand, 87
Mapping, thermal, 6, 122
Mass, conversion factors for, 202
Materials:
  for friction, 86–88
  properties of, 4
  selecting, 5
  for snow fences, 73–74
Materials status as decision-making factor, 131
Measurements of friction, 81–84
Mechanical agitation of snow, 28
Mechanical control methods, 60–61
  for friction, 90–91
  for ice removal, 34
  for snow removal vs. ice removal, 61–62

Media for communication, 147
Melting effectiveness:
  in chemical tables, 245
  of materials, 4
Melting pits, 70
Mercedes Benz Unimogs, 174
Metals, corrosion of, 57–58
Meteorological services, 7
Methyl alcohol, 49
Microwave radiometers, 121
Mineral-insulated cables in heated pavements, 66
Mobile installations for railroad switches, 185–187
Moisture absorption:
  in chemical tables, 246
  of materials, 4
Molecules, water, 30–33
Motor graders, 173–174
Motor patrols for ice control, 110
Motor Truck Engineering Handbook (Finch), 156
MU values for friction, 197

National Weather Service (NWS), 116, 146
Negligence, 137
NEXRAD Doppler radar, 116
Normal force in friction, 78
Nose plows, 97–98
Notice of dangerous conditions, 138
Notice to Airmen (NOTAM), 199
Nozzles, 51–52

Objectives, 133–134
Off-season preparations, 145–146
Off-track railroad equipment, 189
On-track railroad equipment, 188–189
One-way plows, 98–99
Open-loop controllers for spreader spinners, 106–108
Operational strategies, 148–149
Operations plans and manuals, 134–136
Ordinances, 138–139

Parked cars, 139
Particulate matter, 89
Pavements and pavement status:
  in anti-icing strategies, 93
  chemically impregnated, 59–60

Index

Pavements and pavement status (*Cont.*):
  as friction factor, 79
  friction of, 127–128
  heated, 63–67
  ice detectors for, 119–120
  monitoring, 117–121
  surface temperature of, 5–6, 124–125
  thermal mapping of, 122
  (*See also* Roads and road status)
PCC pavements:
  chemical effects on, 9
  sealing, 58
Peak longitudinal braking friction, 84
Performance evaluation, 149–150
Performance requirements in chemical selection, 50–51
Perkins pipes, 65n
Personnel status:
  as decision making factor, 131
  reporting of, 147
Phase diagrams:
  for calcium chloride, 47–48
  for sodium chloride, 44–47
Phase transitions, 33
Phosphate-containing compounds, vegetation damage from, 59
Pilot plows, 182, 188
Pipes in heated pavements, 64–67
Plans and manuals, 134–136
Plastic in heated pavements, 65
Plows, 97
  for airports, 192–193
  disaggregation of snow from, 28
  displacement, 97–102
  and energy loss, 2, 25
  for railroads, 182–184, 188–189
  rotary, 102–103, 172–173, 183–184, 188–189, 193
PlusRide material, 61
Police cooperation, 140–141
Policies, 134–136
  interagency cooperation, 140–142
  legal considerations in, 136–138
  road closures, 139–140
  route priorities in, 139
  snow ordinances in, 138–139
Portland cement concrete:
  chemical effects on, 9
  sealing, 58
Posts for snow fences, 74
Poststorm reporting, 132
Potassium acetate, 243, 253–254

Potassium chloride, 254–255
Powered brooms, 61, 104
Preevent reporting, 131
Preseason preparations, 145
Pressure:
  conversion factors for, 203
  and freezing-point, 201
  in ice, 34
  in ice melting process, 35, 44–46
Prevention, 3–4
Preventive maintenance for trucks, 173
Prewetting chemicals, 8, 52–53
Prime movers:
  alternative, 173–174
  motor trucks, 154–156
Priorities, route, 139
Private weather/climate information sources, 116
Propylene glycol:
  for airports, 195
  characteristics of, 243, 255–256
  cost of, 49
  for heated pavements, 65
Prussian blue, 221
Public communications, 129–130, 147
Pumping in brine containment, 57

Quartz for friction, 87

Radar, 93, 116
Radiation cooling, ice formation from, 36–37
Radiation in thermal methods, 63, 67–68
Radio stations for communication, 147
Radiometers, 6, 121
Railroad snow and ice control, 181–184
  for electrified systems, 190
  for highway grade crossings, 187
  living snow fences for, 74
  for main lines, 184–185
  off-track equipment for, 189
  on-track equipment for, 188–189
  snow removal equipment for, 95–96
  for switches, 185–187
  for tunnels and snow sheds, 187–188
  for yards and terminals, 185
Rain:
  freezing, 35
  salt loss from, 54
Ramps, airport, 195

Rating systems for performance evaluation, 150
Real-time weather information, 6–7
Rear-axle ratio options for snowplow trucks, 166
Rebars, 9
Record keeping, 147–148
Refresher training, 144–145
Registry of Toxic Effects of Chemical Substances (RTECS), 245
Reliability of trucks, 173
Relocated precipitation in snow fence design, 72
Remote processing units (RPUs) in RWIS systems, 118
Reporting:
  at airports, 197–199
  in decision making, 131–132
Resistance cables in heated pavements, 66
Resistance to bending moment of frames, 159–160
Reversible plows:
  for airports, 192–193
  types of, 99–100
Revolving Snow Shovel, 183
Rime ice, 37–38
Road weather information systems (RWIS), 117
  for anti-icing, 93
  communications facilities in, 118–121
  sensors in, 117–121
  signal processing in, 118
Roads and road status:
  closure policies, 139–140
  in decision making, 128–129, 131
  importance of knowing, 5–6
  as material selection criterion, 5
  (See also Pavements and pavement status)
Roadway maintenance:
  abrasives for, 85–89
  chemical methods, 90–93
  friction in, 77–79
  friction measurements in, 81–84
  friction value in, 84–85
  mechanical methods, 90–91
  slip ratio in, 80–81
  thermal methods, 93
  traction in, 77
Rock salt, 218
Rolba plows, 189
Roller distributors for spreaders, 106–108

Rollover plows:
  for airports, 193
  operation of, 99–100
Roofed structures for salt storage, 54–56
Rotary plows:
  for airports, 193
  carriers for, 172–173
  operation of, 102–103
  for railroads, 183–184, 188–189
Route priorities, 139
Rubber edges for ice-cutting blades, 110
Rubit material, 61
Runoff containment for salt, 57
Runway Condition Reports, 197
Runways, airport:
  chemical application rates for, 204
  chemical controls on, 195
  friction testing of, 197–199
Russell snowplows, 183

Salt, 41–42
  economics of, 90–91
  eutectic temperature of, 9, 47
  historical use of, 42–43
  runoff containment from, 57
  in sand mixtures, 89
  storage and handling of, 54–57
  vapor pressure changes from, 44–46
  (See also Calcium chloride; Sodium chloride)
Salt-tolerant plants, 59, 222–239
Sand, 42
  for friction, 85, 87
  storage of, 89
Satellite imagery, 93
Saturation point of air, 35
Sealing pavement surfaces, 9, 58
Seasonal variations, awareness of, 2
Selection:
  of chemicals, 48–51
  of materials, 5
Sensors, 6–7, 117–121
Sewer disposal, 70
Side-mounted plows, 101, 171
Sidewalk snow removal, 112
Signal processing in RWIS systems, 118
Signs, warning, 144
Silos for salt storage, 55–56
Single-element rotary plows, 102–103
Single-track railroad plows, 188
Skid performance coefficient (SPC), 83

Index

Skid resistance, 82–83
Slip ratio, 80–81
Slush, 25–26
Snow and snow control:
  adhesiveness of, 26–27
  airport (*see* Airport snow and ice control)
  bonding by, 6
  characteristics of, 11, 17–18
  cohesiveness of, 26
  compressibility of, 23–25
  costs of, 11–12
  density of, 21–22
  disposal of, 70, 149
  drifts, 70–76
  equipment for, 95–103
  hardness of, 22–24, 27–28
  historical amounts, 12–16
  mechanical agitation of, 28
  molecular structure of, 31, 33
  policies for (*see* Policies)
  principles of, 1–7
  railroad (*see* Railroad snow and ice control)
  for sidewalks, 112
  slush, 25–26
  snow crystal formation, 18–21
  temperature instability of, 27
  thermal properties of, 28–30
  uniqueness of, 30
  vehicles for (*see* Snowplow trucks; Trucks; Vehicles)
Snow clouds, 144
Snow desks at airports, 199
Snow dozers for railroads, 184
Snow Fence Guide, 71
Snow fences, 3–4
  design factors in, 71–72
  living, 74–76
  materials for, 73–74
  placement of, 72–73
Snow loaders, 112–113
Snow melters, 70
Snow ordinances, 138–139
Snow removal equipment, 95
  evolution of, 96–97
  plows, 97–103
  power brooms, 104
Snow rollers, 96, 181
Snow sheds, 12, 16, 187–188
Snowblowers:
  for airports, 193

Snowblowers (*Cont.*):
  carriers for, 172–173
  operation of, 102–103
  for railroads, 183–184, 188–189
Snowbreak forests, 74
Snowplow trucks, 156
  axles, 160–163
  bodies, 169–170
  brakes, 166–167
  cabs, 167–168
  engines, 164–165
  frames, 156–160
  hydraulic systems, 169
  lighting systems, 168–169
  transmissions and drivelines, 165–166
  wheels and tires, 163–164
Sodium, environmental effects of, 9
Sodium acetate, 243, 256–257
Sodium chloride, 218
  additives for, 221
  characteristics of, 257–258
  corrosion from, 58
  environmental effects of, 222
  and friction, 85
  gradations of, 219
  mixing concentrations for, 219–221
  phase diagram for, 44–47
  for prewetting chemicals, 52
  salt-tolerant plants for, 222–239
  solubility of, 45–47
  sources of, 218–219
  standards for, 219
  (*See also* Salt)
Sodium formate:
  characteristics of, 243–244, 258–259
  corrosion from, 58
Solar salt, 218
Solid chemicals:
  application of, 51
  selection of, 49
  spreaders for, 104–109
Solubility:
  in chemical tables, 245
  of sodium chloride, 45–47
Solution salt, 218
Solvay process, 212
Spalling, 58
Specific gravity for salt concentrations, 220
Specific heat of snow, 28–29
Speed and plow energy loss, 25
Spinners, 51, 105–108

Spreaders:
    for airports, 193–194
    calibration of, 108
    distributors for, 106–108
    dump boxes, 106
    hopper bodies, 106
    for ice control, 104–105
    for liquid chemicals, 109
    for railroads, 189
    zero-velocity, 51, 108–109
Standard of care, 137–138
Statistical data, collecting, 132
Steel:
    chemical effects on, 9
    for truck frames, 159
Steering-axle tires, 163
Stopping distance in friction measurements, 82–83
Storage and handling of materials, 5
    antiskid, 89
    chemicals, 53–57, 246
Strength of snow, 22–24
Submerged combustion, 70
Supercooled water droplets, 36
Supersaturated air, 35–36
Surface effects:
    in chemical tables, 246
    of materials, 4
Surfaces, pavement (see Pavements and pavement status)
Switches, railroad, 185–187
Swivel plows, 99

Taxiways, airport, 195
Telephone networks, 146
Television stations for communication, 147
Temperature:
    Celsius to Fahrenheit conversion, 205–206
    and chemical performance, 8–9
    in decision making, 124
    eutectic, 4, 9, 47, 245
    in heated pavements, 67
    and ice, 34
    and snow bonding, 6
    and snow compressibility, 25
    and snow hardness, 22–23
    wind-chill chart, 207
Temperature instability of snow, 27
Terminals, railroad, 185
Terms, glossary of, 269–277

Thermal conductivity:
    conversion factors for, 202
    of snow, 29–30
    in thermal methods, 63
Thermal energy in ice melting process, 44
Thermal gradients in depth hoar, 27–28
Thermal mapping, 6, 122
Thermal methods, 62–63
    design factors in, 68–70
    for friction, 93
    for heated pavements, 63–67
    for railroad switches, 185–186
Thermal properties of snow, 28–30
Thermal stresses in heated pavements, 65
Thermistors in RWIS systems, 119
Thermography, 122
Thermometers, 124
Thomson, J. J., 35
Tierney, Charles W., 183
Time of day and season as decision-making factor, 126–127
Tire/pavement friction in roadway maintenance, 77
Tires:
    as friction factor, 79
    for snowplow trucks, 163–164
Torque ratings of snowplow truck engines, 164–165
Tort liability, 136–137
Toxicity:
    in chemical tables, 246
    of materials, 5
Traction, roadway, 77
    (See also Friction)
Trailing plows, 102
Training, 140, 144–145
Transfer case designs for snowplow trucks, 161–162
Transmissions for snowplow trucks, 165–166
Trees for snow fences, 74
Trends, 150–151
Trolley cars, 96
Truckload prewetting in chemical application, 52
Trucks, 153–154
    attachment costs, 171–172
    hitches for, 170–171
    prime movers, 154–156
    reliability and maintainability of, 173
    rotary plows, 172–173
    in snow disposal, 70

Index

Trucks (*Cont.*):
  snowplow (*see* Snowplow trucks)
  specifications for, 175–180
  wing mounts for, 171
Tungsten carbide inserts for ice-cutting blades, 110
Tunnels, railroad, 187
Turnouts, railroad, 185–187
Twin-spinner spreaders, 193–194
Two-element rotary plows, 102–103
Type I and II fluids, 195

Underbody blades for ice control, 110
Underbody plows, 101–102
Unimogs, 174
Uniqueness of snow, 30
Urea:
  for airports, 195
  characteristics of, 241–242, 260–261
  corrosion from, 58
  environmental effects of, 9, 242
  vegetation damage from, 58–59

V-blade plows, 100–101
V-boxes, 106
Vapor pressure in ice melting process, 44–46
Vaporproof containers, 211
Vegetation:
  chemical effects on, 58–59
  salt-tolerant, 222–239
  for snow fences, 74–76
Vehicles, 110–111
  corrosion costs of, 12
  as friction factors, 79
  mechanical requirements for, 111
  visibility for, 111–112
  (*See also* Snowplow trucks; Trucks)
Verglimit product, 59
Visibility:
  meters for, 120–121
  for vehicles, 111–112
Volume:
  conversion factors for, 202
  vs. density, 21–22

Warning lights for snowplow trucks, 168
Warning signs, 144
Water:
  characteristics of, 30–33
  in heated pavements, 64–65
Weather Channel, 146
Weather/climate information, 7
  for anti-icing, 93
  in decision making, 128–129, 131
  plans for obtaining, 146
  radiometers for, 121
  RWIS, 117–121
  sensors for, 6–7
  sources of, 115–117
  thermography, 122
Wedge plows, 184, 188
Weight, conversion factors for, 202
Weight percent of chemicals calculations, 201
Wheel slip in friction measurements, 83–84
Wheels for snowplow trucks, 163–164
Wicks in heat pipe, 65–66
Wind:
  in airport snow and ice control, 192
  conversion factors for, 205
  as decision making factor, 126
  disaggregation of snow from, 28
  in rime ice formation, 38
  RWIS sensors for, 120
  as snow fence design factor, 71–72
Wind-chill temperature chart, 207
Windows, clear-sight, 112
Windshields:
  damage to, 88
  heated, 111
Wing mounts for trucks, 171
Wing plows, 101
Wiper blades, heated, 111
Wyoming truss-type snow fences, 73

Yards, railroad, 185
Yellow prussiate of soda, 221

Zero-velocity spreaders, 51, 108–109

About the Author

L. David Minsk, after 34 years in the U.S. Army Cold Regions Research and Engineering Lab (CRREL), is currently a consultant for CRREL, other government agencies, and several industrial clients.